International Conference on Materials Research and Innovation (ICMARI 2018)

IOP Conference Series: Materials Science and Engineering
Volume 526

Bangkok, Thailand
17 - 21 December 2018

Editors:

Wirasak Smitthipong
Saree Phongphanphanee
Jirasak Wong-ekkabut

ISBN: 978-1-5108-9808-0
ISSN: 1757-8981

Printed from e-media with permission by:

Curran Associates, Inc.
57 Morehouse Lane
Red Hook, NY 12571

Some format issues inherent in the e-media version may also appear in this print version.

Institute of Physics
Dirac House, Temple Back
Bristol BS1 6BE UK

Phone: 44 1 17 929 7481
Fax: 44 1 17 920 0979

techtracking@iop.org

Additional copies of this publication are available from:

Curran Associates, Inc.
57 Morehouse Lane
Red Hook, NY 12571 USA
Phone: 845-758-0400
Fax: 845-758-2633
Email: curran@proceedings.com
Web: www.proceedings.com

Table of contents

Volume 526

The International Conference on Materials Research and Innovation

17–21 December 2018, Bangkok, Thailand

Accepted papers received: 3 April 2019
Published online: 8 August 2019

Preface

Papers

IOP Conf. Series: Materials Science and Engineering **526** (2019) 011001 doi:10.1088/1757-899X/526/1/011001

IOP PROCEEDING

Organized by

1. Specialized center of Rubber and Polymer Materials in agriculture and industry (RPM), Faculty of Science, Kasetsart University.
2. The Thailand Research Fund.

Sessions

1. Rubbers and Composites
2. Biomaterials
3. Materials of Energy and Environmental Applications
4. Computational Model and Simulations
5. Industrial Innovations

Series Seminar for FREE

"Rubber Science and Innovation"
speaker form The University of Akron, USA.

Dr. Li Jia
(Associate Professor of Polymer Science)

Dr. Sadhan C. Jana
(Professor of Polymer Engineering)

Sponsored by

IOP Conf. Series: Materials Science and Engineering **526** (2019) 011001 doi:10.1088/1757-899X/526/1/011001

Preface

The International Conference on Materials Research and Innovation (ICMARI) is organized on 17th – 21st December 2018 in Bangkok, Thailand. The conference is conducted in frontier research on materials research and innovation including Rubbers and Composites, Biomaterials, Materials of Energy and Environmental Applications, Computational Model and Simulations, and Industrial Innovation. The ICMARI was announced in order to create and build up collaboration network and share experience among researchers in ASEAN and around the world. It is a great opportunity for all participants to exchange knowledge and strengthen the research collaboration. We hope that the participants will be fruitful by invited speakers in knowledge, application, novel technology in the field of materials research and innovation.

Thank you all for participating,

ICMARI 2018 Organizing Committee

IOP Conf. Series: Materials Science and Engineering **526** (2019) 011002 doi:10.1088/1757-899X/526/1/011002

International Conference on Materials Research and Innovation ICMARI

Editors of Proceeding

Dr. Wirasak Smitthipong	Chair Editor
Dr. Saree Phongphanphanee	Editor
Assoc. Prof. Jirasak Wong-ekkabut	Editor

Organizing Committee (Kasetsart University, Thailand)

Dr. Wirasak Smitthipong	Chairman
Dr. Saree Phongphanphanee	Committee
Dr. Nattasamon Petchsang	Committee
Dr. Chanapa Kongmark	Committee
Dr. Decha Dechtrirat	Committee
Dr. Pongthep Prajongtat	Committee
Assoc. Prof. Sutee Boonchui	Committee
Assoc. Prof. Jirasak Wong-ekkabut	Committee
Dr. Chomdao Sinthuvanich	Committee
Asst. Prof. Wirunya Keawwattana	Committee
Asst. Prof. Kiadtisak Saenboonruang	Committee
Assoc. Prof. Patthra Pengthamkeerati	Committee
Asst. Prof. Jukkrit Mahujchariyawong	Committee
Assoc. Prof. Nanthiya Hansupalak	Committee
Asst. Prof. Prakit Sukyai	Committee
Asst. Prof. Parames Kamhangrittirong	Committee
Ms. Ampaiwan Taweetanyaluk	Committee
Ms. Nipasiri Tapchai	Committee
Ms. Ratchadathip Purnajoti	Secretary
Ms. Sukontip Suethao	Secretary
Ms. Thridsawan Prasopdee	Secretary

International Technical Committee

Dr. Wirasak Smitthipong	Kasetsart University	Thailand
Prof. Sadhan C. Jana	University of Akron	USA
Dr. Arkadiusz Chworos	Polish Academy of Sciences	Poland
Prof. Katsura Nishiyama	Meijo University	Japan
Assoc. Prof. Li Jia	University of Akron	USA
Prof. Mikko Karttunen	Western University	Canada
Prof. Koichiro Mitsuke	Josai University	Japan
Assoc. Prof. Norio Yoshida	Kyushu University	Japan
Dr. Saree Phongphanphanee	Kasetsart University	Thailand
Prof. Takaomi Kobayashi	Nagaoka University of Technology	Japan
Dr. Nattasamon Petchsang	Kasetsart University	Thailand
Dr. Jiye Chen	University of Portsmouth	UK
Assoc. Prof. Sutee Boonchui	Kasetsart University	Thailand

IOP Conf. Series: Materials Science and Engineering **526** (2019) 011002 doi:10.1088/1757-899X/526/1/011002

International Conference on
Materials Research and Innovation ICMARI

IOP Conf. Series: Materials Science and Engineering **526** (2019) 011003 doi:10.1088/1757-899X/526/1/011003

Peer review statement

All papers published in this volume of *IOP Conference Series: Materials Science and Engineering* have been peer reviewed through processes administered by the proceedings Editors. Reviews were conducted by expert referees to the professional and scientific standards expected of a proceedings journal published by IOP Publishing.

IOP Conf. Series: Materials Science and Engineering **526** (2019) 012001 doi:10.1088/1757-899X/526/1/012001

Effect of tackifier resin and non-rubber components on adhesive property of natural latex

T Prasopdee[1], K Sriroth[2] and W Smitthipong[1,3,*]

[1] Specialized center of Rubber and Polymer Materials in agriculture and industry (RPM), Department of Materials Science, Faculty of Science, Kasetsart University, Bangkok, Thailand.
[2] Department of Biotechnology, Faculty of Agro-Industry, Kasetsart University, Bangkok, Thailand.
[3] Office of natural rubber research program, The Thailand Research Fund, Bangkok, Thailand.

*Corresponding author: fsciwssm@ku.ac.th

Abstract. The objective of this study is to better understand the adhesive property of natural latex using phenolic resin at different ratios. Whole natural latex (WNL) and deproteinized natural latex (DPNL) are used. By increasing the resin ratio, the total solid contents (TSC) and viscosity for both types of latex increase. Interestingly, DPNL presents better adhesive properties than WNL for a given resin ratio. The phenolic resin at 3 phr gives the best adhesive property of latex. Beyond this resin ratio, the adhesive property decreases. Nevertheless, the phenolic resin can foster the flow properties in latex during the aging time, thus increasing the adhesive property of latex.

1. Introduction

Natural latex from rubber tree (*Hevea brasiliensis*) looks like a milky white colloid, it's dry rubber content or *cis*-1,4-polyisoprene is about 30-40% floating in the colloid. There are also other components besides rubber present in latex, we call these non-rubber components. So, natural latex consists of 5-6% non-rubber components such as proteins, lipids, carbohydrates, minerals, metal, etc. The effect of such non-rubber constituents on some natural rubber properties have been reported [1-2]. In particular, the non-rubber components can greatly affect the adhesive property of natural rubber. From a practical point of view, this property is of great importance for the natural rubber is frequently used as a matrix in composite materials, for example tires, belts and coated fabrics, or in contact with other solids. It is well-known that natural latex can be used as general glue [3-4], however, the quality of glue from natural latex is not stable due to its non-rubber constituents.

Recently, many chemicals have been added to natural latex to make glue. Most of the formulas are secretive, expensive and have a strong latex smell. When applied to products such as handbags, it can cause allergic reactions in humans. Normally, a fast and convenient way to increase the adhesive property of natural latex is to add the tackifier resin which has many types, such as epoxy resin, vinyl resin, phenolic resin, etc [5-6]. However, epoxy resin and vinyl resin may cause allergic reactions [7-8] but the phenolic resin has no reports of allergic reactions. The phenolic resin is inexpensive and water-based which is compatible with natural latex. The objective of this study is to find the effect of the phenolic resin and non-rubber components on the adhesive property of natural latex. Two types of natural latex were prepared: whole natural latex (WNL) and deproteinized natural latex (DPNL). The purified natural latex can, more user-friendly, remove the allergen proteins, as more user-friendly.

2. Experimental

WNL and DPNL were prepared from *Hevea* latex supplied from Thailand, according to a well-defined method [2]. The formulas of adhesive WNL and adhesive DPNL are presented in the table 1 (PR means phenolic resin, KL means 20%wt potassium laurate and KOH means 10%wt potassium hydroxide).

Table 1. Formulas of adhesive WNL and adhesive DPNL in phr (parts per hundred of rubber).

Formula	C1	B1	B2	B3	B4	B5	C2	B6	B7	B8	B9	B10
WNL	100	100	100	100	100	100	-	-	-	-	-	-
DPNL	-	-	-	-	-	-	100	100	100	100	100	100
PR	-	0.6	1.8	3	6	9	-	0.6	1.8	3	6	9
KL	3	3	3	3	3	3	3	3	3	3	3	3
KOH	3	3	3	3	3	3	3	3	3	3	3	3

FTIR (Nicolet Magna 850) in Attenuated Total Reflectance (ATR) mode with diamond crystal was used to determine the chemical structure of the rubbers. Nitrogen content of latex samples (WNL and DPNL) was determined using the Kjeldahl method [9]. Total solid content (TSC) of natural latex adhesive were measured following ISO 124. Viscosity of latex was determined by Brookfield Viscometer (LDVD-III, Brookfield, USA) type rational viscometer. The surface energies of the different rubbers and fabrics were evaluated from contact angle measurements using different liquid drops of known properties [2]. Average separation force of two nylon fabrics adhered by adhesive natural latex were recorded on a tensile testing machine (Instron) type peeling test 180° following ASTM 638. Scanning electron microscope or SEM (XL30 Philips, Netherlands) was used to investigate the topology of the fractured surface between the two fabrics adhered by natural latex adhesive. Then, the fractured surfaces were sputter-coated directly with Au/Pd to observe the surface of fabrics. JEOL JSM-5310 at 10kV acceleration voltage.

3. Results and discussion

First of all, WNL and DPNL were cast as a film in order to analyse the presence of non-rubber constituent for both types of latex using FTIR-ATR (spectra are not shown here). It showed clearly that the DPNL has lower intensity of the protein peaks (Amide I at 1660-1630 cm^{-1} and Amide II at 1575-1480 cm^{-1}) than WNL. This means that WNL contained the protein components, and the contents of these non-rubber constituents were substantially reduced in purification treatment. This result is in good agreement with the previous result [2]. The removal of such non-rubber constituents is important given that some people are allergic to proteins in whole natural rubber. Table 2 shows the measurements of nitrogen content, pH, TSC and viscosity in both WNL and DPNL samples. We found that the quantity of nitrogen content in DPNL is reduced by 82% compared to that of WNL, indicating the protein content can be reduced after the purification of rubber. This establishes confidence in the centrifugation method used here to purify the natural rubber [10]. This result is in good agreement with that of FTIR-ATR. There are no significant differences of the pH, TSC and viscosity between two types of latex.

Table 2. General properties of WNL and DPNL

Sample	Nitrogen (%)	pH	Viscosity (cP)	TSC (%)
WNL	0.76 ± 0.08	10.41	7.5	37.07 ± 0.05
DPNL	0.14 ± 0.08	11.29	7.5	37.20 ± 0.14

Concerning to the characterization of adhesive WNL and adhesive DPNL, we found that adhesive WNL samples (formulas C1 to B5) have similar pH values (during 10.3 - 10.4), whatever the quantity of phenolic resin, similar results with those of adhesive DPNL samples (formulas C2 to B10). However, the viscosity value and TSC trend to increase with the increasing of phenolic resin's concentration, especially beyond 3 phr of phenolic resin in both adhesive WNL and adhesive DPNL.

The surface energy of natural rubbers and fabric can be determined using contact angle measurements. It is well-known that a better knowledge of the adhesion phenomena is required for practical applications of multi-component materials. Moreover, it is considered that adhesion between two solids is due to interatomic and intermolecular forces established at the interface, provided that an intimate contact is achieved [11]. The most common interfacial forces result from van der Waals and Lewis acid–base interactions. The magnitude of these forces can generally be related to fundamental thermodynamic quantities, such as surface free energies of both entities in contact [12]. We found that both types of natural rubber films represent as a non-polar material (below 50 mJ/m^2) and nylon fabric represents as a polar material (above 50 mJ/m^2), in good agreement with the previous work [13]. Then, the work of adhesion (W) between rubber and nylon fabric can be calculated [13]. So, the work of adhesion for the two types of adhesive latex with nylon fabric is fairly similar for all the samples; this may not explain the adhesive property in WNL/nylon (W = 92.9 mJ/m^2) and DPNL/nylon (W = 91.2 mJ/m^2) samples.

The adhesive property of adhesive latex (WNL and DPNL) on nylon fabric can be assessed by the average separation force in the same contact area, better adhesive property of adhesive latex does require above average separation force to separate two nylon fabrics. Figure 1 shows the average separation force as a function of adhesive latex formula. The adhesive property of both adhesive latex trends to increase with the increase of phenolic resin concentration until 3 phr (formulas B3 and B8). Beyond this concentration, the separation force decreased. From the topological point of view (SEM images), we can find the insoluble resin beyond 3 phr of resin on the surface of fabric after performing the 180° peeling test, which may cause the decrease in adhesive property of adhesive latex.

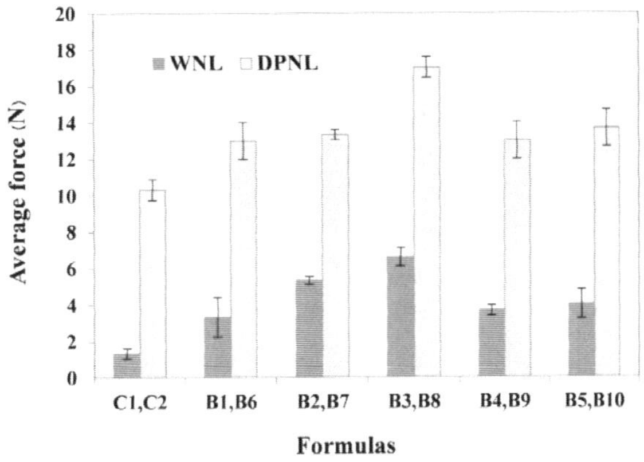

Figure 1. Average force for 180° peeling test of nylon fabric coated with samples (formulas C1, B1 to B5 for adhesive WNL and C2, B6 to B10 for adhesive DPNL).

However, adhesive DPNL has higher average separation force than adhesive WNL in the nylon fabric. For a given phenolic concentration and for both types of adhesive latex, the dependence of adhesive property on the nature of latex is observed. The presence of non-rubber components plays a major role in the adhesive property, since the adhesive WNL always presents the weakest adhesive property and the extracted natural latex (adhesive DPNL) always leads to the strongest adhesive property. The reduction of protein components leads, therefore, to an increase of adhesive property. This result indicates a strong dependence of adhesive property on the presence of these components in the natural latex. The possible explanation lies in the presence of proteins within natural latex and, thus, at the interface, affecting the magnitudes of adhesive property. This result is in good agreement with the previous works [2, 10] that migration of low mass components at the surface of natural rubber had occurred. By cleaning with acetone, this surface layer could be removed, resulting in large increase of the adhesion energy of the natural rubber.

4. Conclusions

In this research, we studied the effect of tackifier resin as phenolic resin and non-rubber components in natural latex in order to better understand the adhesive property of adhesive latex. Two types of natural latex were used: whole natural latex (WNL) and deproteinized natural latex (DPNL). We found that nitrogen content in DPNL decreased by 82% compared to that of WNL, while other characteristics were similar to WNL. The increasing of tackifier resin increases the total solid contents and viscosity, thus improves the adhesive property of adhesive latex. Because the tackifier resin could help the flow property in latex during aging time, thus increasing the adhesive property of latex. The adhesive property of adhesive DPNR is better than adhesive WNR due to the removal of non-rubber components which affects the quality of latex. This present research work shows the potential of developing eco-friendly natural adhesive which contains stable quality and less allergen protein.

5. References

[1] Smitthipong W, Chollakup R and Nardin M 2014 (Eds.) In Bio-based composites for high-performance materials, *CRC Press*, 324 pages.
[2] Smitthipong W, Nardin M, Schultz J, Nipithakul T and Suchiva K 2004 *J. Adhesion Sci. Technol.* 18 1449.
[3] Chumsamrong P and Monprasit O 2007 *Suranaree J. Sci. Technol.* **14** 269.
[4] Smitthipong W, Nardin M, Schultz J and Suchiva K 2007 *Inter. J. Adhesion Adhesives* **27** 352.
[5] Mess A, Vietzke J-P, Rapp C and Francke W 2011 *Anal. Chem.* **83** 7323.
[6] Sham A and Krishna M 2011 *J. Eng. Res. Stu.* 62.
[7] http://allergies.about.com/od/contactdermatitis/a/Allergy-To-Bandages-And-Adhesives.htm: July 2018
[8] http://dermnetnz.org/dermatitis/epoxy-allergy.html; July 2018.
[9] Sáez-Plaza P, Michałowski T, José Navas M, García Asuero A and Wybraniec S 2013 *J. Crit. Rev. Anal. Chem.* **43** 178.
[10] Smitthipong W, Tantatherdtam R, Kanokwan R, Suwanruji P, Sriroth K, Radabutra S, Thanawan S, Vallat M-F, Nardin M, Mougin K and Chollakup R 2014 *Adv. Mater. Res.* **844** 345.
[11] Shah DU, Porter D and Vollrath F 2014 *Compos. Sci. Technol.* **101** 173.
[12] Shayestehfar S, Yazdanshenas ME, Khajavi R and Rashidi A-S 2014 *J. Eng. Fiber Fabr.* **9** 158.
[13] Smitthipong W, Suethao S, Shah D and Vollrath F 2016 *Polym. Testing* **55** 17.

Acknowledgement

This work was funded by Kasetsart University Research and Development Institute (KURDI Grant number 26.57), Kasetsart University, Thailand.

IOP Conf. Series: Materials Science and Engineering **526** (2019) 012002 doi:10.1088/1757-899X/526/1/012002

FEM-based assessment of wear of stamping die

T Trzepiecinski[1,*] **and H G Lemu**[2]

[1] Rzeszow University of Technology, Department of Materials Forming and Processing, al. Powst. Warszawy 12, 35-959 Rzeszów, Poland
[2] University of Stavanger, Department of Mechanical and Structural Engineering and Materials Science, N-4036 Stavanger, Norway

[*] Corresponding author: tomtrz@prz.edu.pl

Abstract. In sheet metal forming industry, surface damage and wear such as adhesion and ploughing of metallic sheet, is a well-known problem which reduces the tool life and quality of drawn elements. Surface defects are the consequences of wear and they accelerate the galling process. The knowledge of places particularly endanger wear allows appropriate design of tools. In this paper, an investigation that focused on the finite element-based analysis of wear of stamping tool for forming an axisymmetric drawpiece has been reported. The analyses were carried out for deep-drawing quality steel sheet with a sheet thickness of 2 mm. The basic mechanical parameters of the sheet tested are determined in uniaxial tensile tests. The analyses performed using Marc/Mentat program for nonlinear analyses take into account the strain hardening phenomenon of sheet material and different values of friction coefficient. The implementation of an Archard's wear model in the numerical simulation proved the possibility of tool wear simulation in sheet metal forming. As a result of the conducted tests, the places of the stamping die potentially exposed to quick wear were determined. It was found that the most exposed region on accelerated wear is the upper part of the die radius.

1. Introduction

Tool wear is an indispensable process in the exploitation of tools in metal forming. The term wear is understood as the loss of material from the surface of tools, which occurred under the influence of friction and the accompanying physical or chemical phenomena [1, 2]. Worn-out tools are Often regenerated. Tool surface damage occurs due to faulty construction, wrong manufacturing technology or improper operation. Inappropriate design of tools can be a source of stress accumulation, which under the influence of dynamic loads are sources of cracks. Another parameter determining the tool wear is an inappropriate selection of material with insufficient hardness and, in the case of heat treatment, low heat resistance. Scratches and too much surface roughness of the tools reduce the strength of the material and are also sources of cracks. The basic types of wear include abrasive wear, fatigue and thermal adhesive wear. Abrasive wear is characterized by a set of phenomena occurring in the contact areas of two moving elements relative to each other. Abrasive wear occurs at relative speed and unit pressures resulting in only a slight increase in tool temperature. According to the literature [3], there exist many approaches of wear modeling based on contact mechanics and failure mechanics. The most known is the Archard's model [4] which is based on the motion and interaction of opposing asperities on either contacting body. Archard assumed that the deformation occurring was of a plastic type and that the material property of greatest importance was the "flow pressure" of the softer metal

[5]. Savio et al. [6] analysed the polishing of glass moulds and the inherent material removal process. This is a very complex process because the element will be a polished with the polishing tool and an abrasive action due to the slurry used to help to remove material.

This paper presents the results of numerical investigations of the prediction of places of wear of stamping tool based on the Archard wear model. The wear analyses were carried out during forming of the axisymmetric drawpiece from deep-drawing quality (DDQ) steel sheet. The wear value was determined in the FE-based Marc/Mentat program.

2. Material

The sheet forming tests of axisymmetric drawpiece were conducted on deep drawing quality steel sheet with a sheet thickness of 2 mm. The chemical compositions of steel used are listed in Table 1. The mechanical properties of the sheet metals (Table 2) have been determined through uniaxial tensile test. Five samples were tested and average values of mechanical parameters were determined. The values of the strain hardening parameters K and n in Hollomon law ($\sigma = K \cdot \varepsilon^n$) are determined from the logarythmic true stress – true strain plot by linear regression.

Table 1. Chemical composition of steel sheet (wt. %)

C	Mn	Mo	Ni	Cr	Nb	S	P
0.09	0.18	0.02	0.05	0.05	0.006	0.018	0.04

Table 2. The selected mechanical properties of the tested DDQ sheet*

Material	E (GPa)	$R_{p0.2}$ (MPa)	R_m (MPa)	A_{50} (%)	K (MPa)	n
DDQ	189.4	182.5	304.9	43.1	583.6	0.230

* E – Young's modulus, $R_{p0.2}$, yield stress, R_m – ultimate tensile stress, A_{50} – elongation, K – strain hardening coefficient, n – strain hardening exponent

3. Numerical modeling

FE-modeling of stamping process was carried for the axisymmetric drawpiece. Due to symmetry of the forming process of axisymmetric drawpiece, the model of the forming process has been simplified to 2D (Fig. 1). In real forming conditions, the sheet flange has tendency to loss the stability. To protect the sheet against wrinkling in the blankholder contact region, the force of value PB = 3 kN is applied (Fig. 1). The punch and blank diameters were equal to 95 and 220 mm, respectively. Edges of tools were rounded by the radius of 10 mm. All parts of the stamping die are assumed as deformable.

Due to the fact that the strength of tools is much higher than the strength of the drawpiece, the tools are considered as perfectly elastic with the values of elastic parameters Young's modulus and Poisson's ratio assumed according to the literature [7]: E = 2.07 GPa and ν = 0.3, respectively. The workpiece is considered as elastic-plastic strain strengthenable material with parameters of strain hardening K and n listed in Table 2. The elastic properties of the sheet material are as follows E = 2.045 GPa and ν = 0.3. The blank and tools were modeled with a four-node, isoparametric, arbitrary quadrilateral elements. The mesh is concentrated in the vicinity of surfaces of tools which are in contact with workpiece. The workpiece consists of 880 elements, while tools consist 14020 elements.

The contact phenomena were described by two bilinear Coulomb's and shear friction models. The coefficients of friction between the tools and workpiece were assumed to be μ = 0.1, μ = 0.2 and μ = 0.3 [8]. According to this study, the analysed values of friction coefficients in contact of deep-drawing quality steel sheets correspond to the real values [9].

In order to determine the tool wear, the Archard's abrasive wear model [4] based on the stress in contact and the rate of displacement of the material at workpiece-tools contact was applied. The total volume of wear debris produced w according to Archard's model can be determined from the formulae:

$$w = KF\frac{G}{H} \tag{1}$$

where K is the wear coefficient, F is a normal load, G is the sliding distance and H is the hardness.

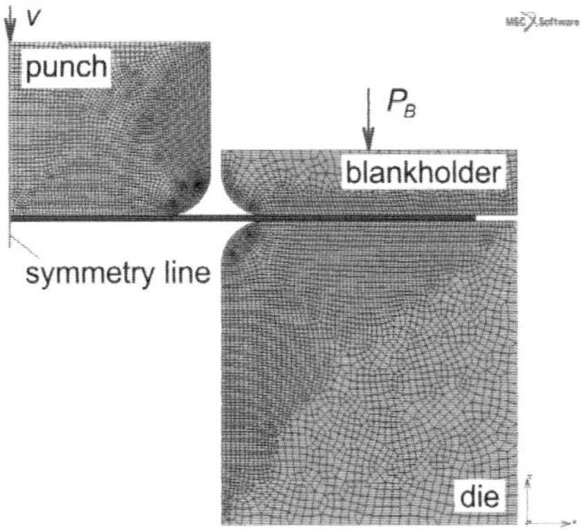

Figure 1. Schematic illustration of the tools and blank for the FE model

After converting Eq. 1 into an incremental form and upon assuming that wear is calculated as a local quantity, the normal force can be replaced by normal stress, and wear rate $\dot{w}$ takes the form:

$$\dot{w} = \sigma\frac{K}{H}v_r \tag{2}$$

where $\dot{w}$ is the rate of change of wear in the direction normal to the surface, σ is the normal stress, v_r is the relative sliding velocity.

The incremental wear is calculated as $\dot{w}\Delta t$ and the wear is accumulated as

$$w_{n+1} = w_n + \dot{w}\Delta t \tag{3}$$

where w_{n+1} is the actual value of wear depth, w_n is the value of wear in previous step of computation, Δt is the time step of computation.
The wear coefficient value and hardness are assumed to be 52 HRC and 0.0178 [8].

4. Results

The prediction of tool wear in the numerical model was analyzed in the area of blankholder action and on the rounded edge of the die. These regions of the stamping die are the most expose to the wear phenomenon. The conditions in the die radius are crucial to the overall tool wear. It is in consistence with the results of Pereira et al. [10]. The wear prediction in both analysed friction conditions is similar. It was found that there are two areas of accelerated wear (Fig. 2). The first region is located on the upper part of the rounded edge of the die. The second region is located at the entrance of the sheet from the blankholder area. For the results to be readable, the distribution of wear presented in Figs 2 and 3 omitted the areas with very little wear below 10^{-5} μm.

Wear in the flat region of the contact zone of sheet with blankholder is relatively small. The force imposed by blankholder is distributed on the very big area of contact of the sheet with blankholder. So, the normal pressures are relatively small.

The change in the curvature of the sheet in region A (Fig. 2) produces an increasing plastic strain of the material and increases the pressure of the sheet material on the tool surface. At rounded edge of the die, the wear is very uniform. The increase in the value of friction coefficient slightly influences the tool wear (Figs. 2, 3). Examination of the wear indices and local stress concentrations for the other radiuses of the tools indicated similar results, i.e. the region of extensive wear and stress concentration is on the rounded edge of the die. According to the FE-based numerical results, the predicted wear locations were consistent with the trend of experimental results available in the literature [10, 11].

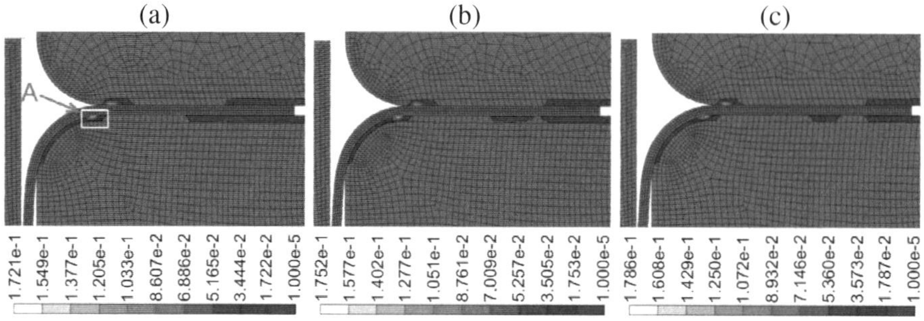

Figure 2. Wear (μm) in the blankholder area for the Coulomb friction conditions at punch depth of 30 mm: (a) $\mu = 0.1$, (b) $\mu = 0.2$ and (c) $\mu = 0.3$.

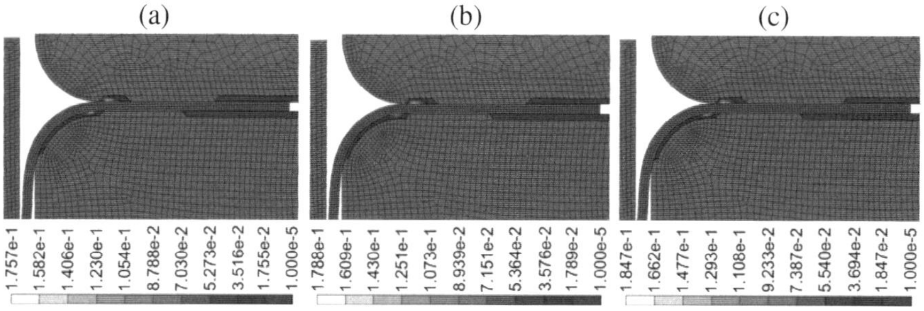

Figure 3. Wear (μm) in the blankholder area for the shear friction conditions at punch depth of 30 mm: (a) $\mu = 0.1$, (b) $\mu = 0.2$ and (c) $\mu = 0.3$.

5. Summary

The implementation of the Archard's wear model in the finite element-based numerical modeling proved the possibility of tool wear simulation in the sheet forming. The possibility of numerically predicting tool wear is helpful in determining the period of reliable work of the stamping die. The methodology presented in the work can be used at the design stage of the tool to determine the places of accelerated wear. In this way, it is possible to use the removable inserts only in places where wear rate is crucial. The results obtained indicate that the upper region of the rounded edge of the die is the most exposed to wear. Furthermore, the change in friction conditions slightly influences the friction coefficient value.

6. References

[1] Jaworski J, Kluz R and Trzepiecinski T 2016 *Eksploat. Niezawodn.* **18** 271
[2] Jaworski J and Trzepiecinski T 2016 *Kovove Mater.* **54** 17
[3] Meng H C and Ludema K C 1995 *Wear* **181-183** 443
[4] Archard J F 1953 *J Appl. Phys.* **24** 981
[5] Montgomery S, Kennedy D and O'Dowd N 2009 *Proc. Int. Conf. on Materials, Tribology, Recycling* (Lipanj: Croatia) pp. 1-17
[6] Savio G, Meneghello R and Concheri G 2009 *Int. J. Mach. Tool. Manu.* **49** 1
[7] Okafar U C 2012 *Mechanical Characterization of A2 and D2 Tool Steels by Indentation*, MSc Thesis (Denton: University of North Texas)
[8] Deng L, Mozgovoy S, Hardell J, Prakash B and Oldenburg M 2017 *Int. J. Mater. Form.* **10** 717.
[9] Trzepieciński T, 2010 *Arch. Civ. Mech. Eng.* **10** 95
[10] Pereira M P, Yan W, Rolfe B F 2012 *Wear* **274-275** 355
[11] Bofer C, Attaf D, Penazzi L and Levaillant C 2005 *Wear* **259** 1097

IOP Conf. Series: Materials Science and Engineering **526** (2019) 012003 doi:10.1088/1757-899X/526/1/012003

Development of rapid qualitative detection DNA microarray platform for *Salmonella typhimurium* for food safety

C Samphaongern[1], C Pipatpanukul[2] and T Srikhirin[1,*]

[1]Materials Science and Engineering Program, School of Materials Science and Innovation, Faculty of Science, Mahidol University, Bangkok 10400, Thailand
[2]Materials Engineering Program, Faculty of Engineering, Burapha University, Chonburi 20131, Thailand

*Corresponding author: toemsak.sri@mahidol.ac.th

Abstract. Screening of possible bacteria contamination in food products is one of the measures implemented to avoid potential health hazards. In food industry, standard cell culture technique is widely used to monitor bacterial contamination. However, the main drawback of this technique is its inherited time consumed during the culturing step, which requires about 72 hours. In this research, an alternative DNA microarray technique was developed for a qualitative screening of a possible *Salmonella typhimurium* contamination in chicken, with the aim of reducing the detection time down to 3 hrs. The identification of *S. typhimurium* in food samples was carried out by hybridization of the possible contaminant with specific probe immobilized on a biochip. Novel DNA probes were designed as 16-50 base pairs of nucleotides to recognize with specific parts in *S. typhimurium* genomic. At least two DNA probes were identified as the candidate probes which had the potential to promote the best hybridization. DNA microarray was fabricated by mixing candidate DNA probes with photoactive polymer network, (poly(DMAA-mABP-SSNa)), and was printed onto a plastic substrate by non-contact microspotter. DNA probe was covalently immobilized onto the surface by 254 nm of UV lamp with 1.25 J/cm2. PCR product functionalized with biotin was hybridized with DNA probe and labelled with streptavidin-cy5. Specific binding yielded fluorescence signal. The intensity image signal was read-out by a fluorescent microarray reader. Two genes specific to *S. typhimurium* (*fimC* and *invA*), were investigated by using specially designed DNA primers and DNA probes. From the assay optimization, it was found that 1 mg/ml of polymer hydrogel concentration, 10 μM of DNA probe concentration, 10 μg/ml of labelling concentration, and 2 nL of array volume yielded the highest signal intensity. The results were calibrated into CFU/ml (cell forming unit). The system was applied successfully for the detection of *S. typhiurium* without any contamination. The biochip validation with spiked sample (DNA standard) is currently underway.

1. Introduction

Food pathogen refers to microorganisms that cause disease in humans and animals by food contamination. The major food pathogen that can lead to foodborne illness is bacteria. Many types of bacteria are concerned with fresh chicken but the one that plays an important role is *Salmonella typhimurium*. Recently, two important genes that are used for identification of *S. typhimurium* such *fimC* and *invA* were reported [1,2]. However, many methods have been recently developed for microbial identification based on cell culture technique in the standard technique, the growth of target cells on a specific media, still commonly used because of its cost effectiveness. Contrarily, the identification process can take up to 24-72 hours, which can be a major drawback, especially, in the food industry.

DNA microarray technology is a novel technique that can be fabricated to contain thousands of the genetic code in a single chip. The technique relies on designing smaller fragments of DNA probes to represent the known genes of target cells. These DNA probes are then spotted on fixed positions on a substrate. DNA microarray technique is an alternative technique for the multiplex detection where the multiplex of DNA probes are being immobilized to surface of substrate. Photoactive polymer network applied to DNA microarray fabrication was proposed by previous work. the method begins with mixing designed DNA probes and synthesized photoactive polymer network, (poly(DMAA-mABP-SSNa)), which are then printed onto a plastic substrate (PMMA sheet) by non-contact microspotter. In a previous study, DNA probe was covalently immobilized onto surface by UV irradiation at 254 nm wavelength with 1.25 J/cm2 energy. PCR product functionalized with biotin were then hybridized with the DNA probe at the annealing temperature and labelled with streptavidin-cy5 [3]. This specific binding yields fluorescence signal and the intensity image signal can be read-out by a fluorescent microarray reader.

In this study, primers and probes as specific to *fimC* and *invA* were designed to amplify DNA product and promote specific hybridization, respectively as shown in table1. However, a step of food pathogen DNA microarray fabrication was optimized to obtain the best intensity signal. Important parameters need to be optimized, including polymer concentration, array volume, labelling concentration, and DNA probe concentration. In this study, DNA array sensor was fabricated for the detection of *S. typhimurium* using probes based in *invA* and *fimC* genes and was tested with DNA standard sample for initiating the platform for real sample testing.

Table1. Target genes, designed primer and designed probe

Target genes	Designed primers (5'-3')	Number of designed probes
fimC	F: CTCTGGCAATCAGTAATAGCGA R: CTGATATTGACCAGCGTGAG	2
invA	F: TTAGGACTGATTGGCGATCTC R: CTCAACTTCAGCAGATACCA	3

2. Materials and Methods

2.1. Optimization of polymer concentration
The photoactive polymer network was synthesized followed by previous study [3]. Concentration of polymer solution was diluted as 0.1, 0.5, 1, 2, 5 mg/ml with distilled water and mixed with 100 µg/ml of labelling solution (streptavidin-Cy5 (GE Healthcare)). Spot volume was 0.4 nL (lowest volume of droplet) which was printed on PMMA substrate (Rungart company, Thailand) by microspotter machine (sciFLEXARRAYER, SCIENION) and incubated in a 254nm-UV chamber (UV Stratalinker 1800, STRATAGENE) of 1.25J. The substrate was washed with PBS buffer (Sigma Aldrich) and observed with a microarray reader (Agilent High-Resolution Microarray Scanner).

2.2. Optimization of microarray spot volume
1mg/ml of polymer solution was mixed with 100 µg/ml of labelling The spot volume was varied as follows 0.4, 1.2, 2.0, 2.8, 3.6 nL and was printed on PMMA substrate and incubated in a 254nm-UV chamber of 1.25J. The substrate was washed with PBS buffer and observed with a microarray reader.

2.3. Optimization of the labelling
1mg/ml of polymer solution mixed with labelling solution that varied in concentration of staining as follows 0.001, 0.01, 0.1, 1, 10 and 100 µg/ml respectively. Spot volume was 2 nL of solution which was printed on PMMA substrate and incubated in a 254nm-UV chamber of 1.25J. The substrate was washed with PBS buffer and observed with a microarray reader.

2.4. Optimization of DNA probe density
1mg/ml of polymer solution mixed with DNA probe-biotin that varied in concentration of probe as follows 0.01, 0.1, 1, 10, and 20 µM. Spot volume was 2 nL which was printed on PMMA substrate

and incubated in 254nm-UV chamber of 1.25J, followed by staining with 10 μg/ml of labelling for 10 minute. The substrate was washed with PBS buffer and observed with a microarray reader.

2.5. Optimization of DNA standard sample detection

Standard culture of *S. typhimurium* was cultured as 24 hours. Bacterial suspensions were prepared and standardized (according to Standard McFarland solution) to $1.5x10^8$, $1.5x10^7$, $1.5x10^6$, $1.5x10^5$, $1.5x10^4$, $1.5x10^3$, and $1.5x10^2$ CFU/ml (cell density) respectively, and bacterial DNAs were extracted from each solution. 1 mg/ml of polymer solution was mixed with 10 μM a designed probe and immobilized onto the PMMA surface by UV irradiation at 254 nm with 1.25 J/cm2. DNA standard PCR product functionalized with biotin was hybridized with the designed probe at annealing temperature and labelled with 10 μg/ml streptavidin-cy5 for 10 minutes. The PMMA substrate was washed with PBS and the results were observed with microarray reader. First platform was designed for *fimC* designed probe testing; the platform was divided to 3 fields including positive field (blue), designed probe field (yellow) and negative field (red), blue field for biotin tagged probe, 1-yellow field for *fimC* designed probe1, 2-yellow field for *fimC* designed probe2, 3-red field for *L. monocytogenes* probe, 4-red field for E. coli probe and 5-red field for bare polymer solution without probe [figure1, a]. Second platform was designed for *invA* designed probe testing; platform was divided to 3 field including positive field (blue), designed probe field (yellow) and negative field (red), blue field for biotin tagged probe, 1-yellow field for *invA* designed probe1, 2-yellow field for *invA* designed probe2, 3-yellow field for *invA* designed probe3, 4-red field for *L. monocytogenes* probe, 5-red field for *E. coli* probe and 6-red field for bare polymer solution without probe [figure1, b].

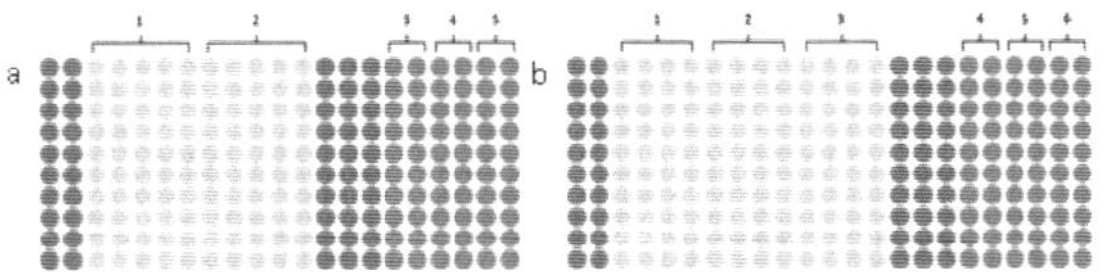

Figure 1. a) *fimC* testing platform b) *invA* testing platform

3. Results and discussion

3.1. DNA microarray fabrication

From results, it found that concentration of polymer solution is 1mg/ml, due to light intensity signal increased from 0.1 mg/ml until it reached a plateau around 1 mg/ml. As the signal did not vary much beyond this point, 1mg/ml of polymer solution was fixed for other experiments [figure2, a]. For the spot volume optimization, there were no significant differences in term of intensity among the spot volumes 1.2, 2.0, and 2.8 nL. From these, 2.0 nl was selected as it had the most uniform morphology and the lowest standard deviation in terms of intensity [figure2, b]. Intensity of labelling conc. trended to increase significantly from 1 μg/ml to 10 μg/ml and attained 85% intensity around 10 μg/ml (97.75% intensity with 100 μg/ml) [figure3, a]. For the limit of reagent, DNA probe density (probe conc.) could be read up to the highest conc. at 20 μM, and also exhibited a stable range around 10 μM [figure3, b]. Following optimization of these parameters, the first platform was fabricated and tested with the standard DNA sample.

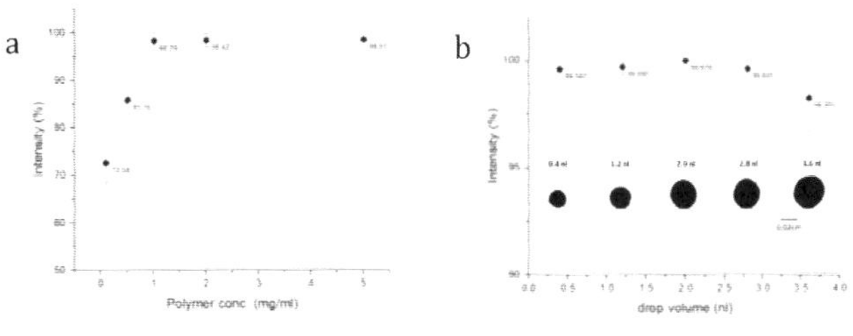

Figure 2. a) effect of polymer solution conc. to signal
b) effect of drop volume to signal

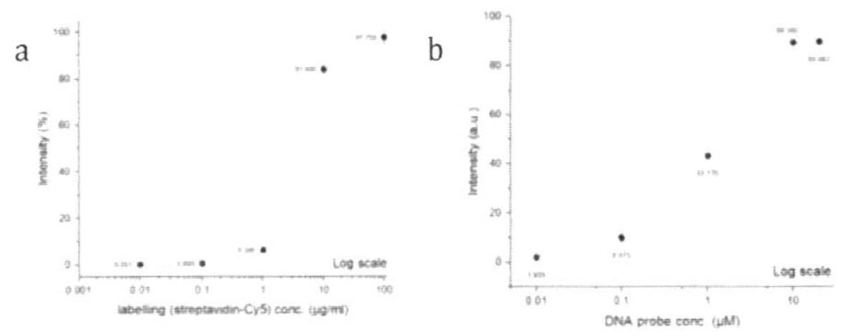

Figure 3. a) signal limitation of labelling
b) effect of DNA probe density to signal

3.2. DNA standard sample testing

For first platform, both designed probes detected DNA product from 1.5x10^{8} CFU/ml until the last concentration of product 1.5x10^{2} CFU/ml, *fimC* probe2 show higher intensity than *fimC* probe1. These results were obtained without contamination with other probes [figure4, a]. In the second platform, only invA probe3 detected DNA product and showed limitation or cut of value at 1.5x10^{4} CFU/ml, without contamination with other probes as well [figure4, b].

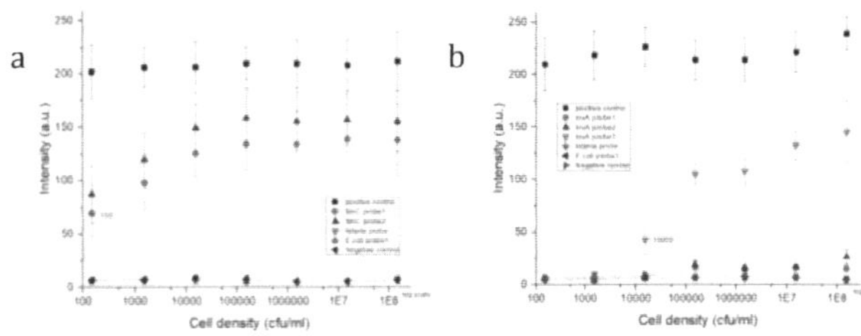

Figure 4. a) detection of DNA sample with *fimC* designed probes b) detection of DNA sample with *invA* designed probes

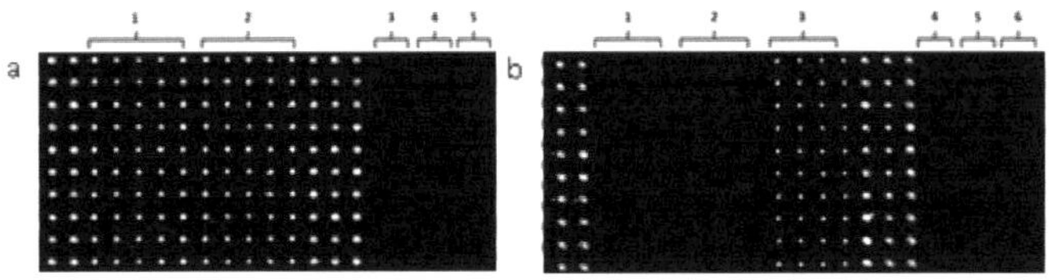

Figure 5. a) *fimC* array at 1.5x10^{7} CFU/ml b) *invA* array at 1.5x10^{6} CFU/ml

4. Conclusions

In this study, important parameters were optimized and comprised of 1 mg/ml of polymer solution, 2 nL of array volume, 10 µg/ml of labeling concentration, and 10 µM of DNA probe concentration for the detection of *S. typhimurium* using DNA micro array platform. With the standard DNA tests, *fimC* gene and *invA* gene could be detected with designed probes and presented a cut off value at 1.5x10^{2} CFU/ml and 1.5x10^{4} CFU/ml, respectively, without any contamination with others probes. The propose platform can qualitatively detect the PCR product of *S. typhimurium*.

5. Reference

[1] Rahn, K., De Grandis, S. A., Clarke, R. C., McEwen, S. A., Galan, J. E., Ginocchio, C., ... & Gyles, C. L. (1992). Amplification of an invA gene sequence of Salmonella typhimurium by polymerase chain reaction as a specific method of detection of Salmonella. Molecular and cellular probes, 6(4), 271-279.

[2] Gonzalez-Escobedo, G., & Gunn, J. S. (2013). Identification of Salmonella enterica Serovar Typhimurium Genes Regulated during Biofilm Formation on Cholesterol Gallstone Surfaces. Infection and Immunity, 81(10), 3770-3780. doi:10.1128/iai.00647-13

[3] Rendl, M., Bönisch, A., Mader, A., Schuh, K., Prucker, O., Brandstetter, T., & Rühe, J. (2011). Simple one-step process for immobilization of biomolecules on polymer substrates based on surface-attached polymer networks. Langmuir, 27(10), 6116-6123.

Acknowledgments

This part of research was kindly supported by Prof. Dr. Jürgen Rühe, Department of Microsystems Engineering, Laboratory for Chemistry and Physics of Interfaces, University of Freiburg - IMTEK, D-79110 Freiburg, Germany, the German Academic Exchange Service (DAAD). Center of Excellence on Medical Biotechnology (CEMB).

Business Process Analysis and Improvement for a Third Party Logistics Provider in Indonesian Cold Chain Logistics

S P Pradita[1] **and P Ongkunaruk**[1,*]

[1]*Department of Agro-Industrial Technology, Faculty of Agro-Industry, Kasetsart University, 50 Ngam Wong Wan Rd. Lad Yao, Chatuchak, Bangkok 10900, Thailand*

* Corresponding author: pornthipa.o@ku.ac.th

Abstract. Indonesian consumer awareness to seek for quality fresh food products is increasing. Therefore, a cold chain logistics is required to deliver agricultural and fishery products across islands. The number of logistics providers in Indonesia are continuously growing, especially for cargo services. Our objectives are to identify the business processes and related activities in a cold chain logistics provider, to analyse problems and to provide recommendations for improvement. We conducted an in-depth interview and analysed the business process using Integration Definition for Function Modelling (IDEF0). The results showed that the major problems are lack of proper cold chain logistics and knowledge in supply chain management. Then, we recommended to improve efficiency in the major activities. These could be the direction for further development to enhance Indonesian cold chain logistics especially for other third party logistics providers to improve their competitive advantages.

1. Introduction

Indonesia is archipelago comprising of over 17,000 islands with five big economic islands approximately 1,904,569 km^2 of land area. There have been difficulties in accessing and delivering fresh products across the islands. Most agricultural products in Indonesia become waste annually for about 40% due to unsuitable storage, shipping, and poor handling of the products [1]. The centre of the economy and food industry is on the Java island, one of which is Surabaya. However, the products must be distributed throughout Indonesia, especially in Eastern Indonesia within a short time. According to Indonesia Cold Chain Association 2016, the number of logistics providers in Indonesia are still growing, especially for cold chain to deliver agricultural and fishery products to the customers [2]. Quality and safety of food products can be maintained from producers to consumers by using proper temperature control system such as reefer containers. Although there are many logistics service providers in Indonesia, Indonesia's national logistics remains inefficient [3]. Therefore, in this research, we explore a case study of third party logistics provider (3 PL's) located in Eastern Java, Indonesia and its supply chain. Our objectives are to identify the business processes and related activities in a third party logistics provider, to analyse problems and to provide recommendations for improvement.

2. Methodology

In this research, we conduct an in-depth interview with the marketing and operational managers of a third party logistics provider in Surabaya, Eastern Java, Indonesia using a questionnaire as a tool to explore the Indonesian cold chain. The questionnaire is used to understand key activities such as planning, sourcing, delivering, and returning in the organizational and the department levels [4]. Then, we analyse a business process using Integration Definition for Function Modelling (IDEF0). The IDEF0 is a method designed to model the decisions, actions, and activities of an organization, process or system [5]. IDEF0 represents flow of activities in the form of square box with a code at the right corner in order to indicate the rank of the activity such as A1, A2, and A3. There are four types of arrows which identify the input, output, control, and mechanism of activities. First, the input is the factor that drive the activities, represented by a left arrow entering into the activity box. Second, the output is the result from doing the activity, represented by a right arrow leaving the activity box. Third, the control is the standard, regulation, or requirement of the activity, represented by an arrow entering the activity box from the top. Fourth, the mechanism implies the resources such as people, equipment, and/or machine that used to perform the activity, represented by an arrow entering into the bottom of the activity box [6]. In addition, a solid line identifies the current activities and a dashed line shows the improvement that we suggested after analysing the business process in order to improve the efficiency of the third party logistics provider in Indonesian cold chain.

3. Results and Discussion

3.1. Background of a third party logistics provider

A company is a third party logistics provider specialized in the transportation of temperature controlled products, both locally, and internationally. Surabaya's location has a Tanjung Perak port which is the second largest port in Indonesia after Tanjung Priok (Jakarta). All shipping activities from inbound or outbound Java must utilize this port. The company has more than 40 branches spread throughout the archipelago in Indonesia. The company specializes in shipping goods with reefer containers for local coverage between islands in Indonesia. The main commodities handled include frozen fish, seafood, meat, frozen chicken, fresh milk, eggs, vegetables, fruit, and other food or beverage products. The product will be shipped from inside or outside the Java island to various parts of the island in Indonesia. There are about 100 trailers, 200 chassis for both 20" and 40", 1,700 reefer containers units consisting of 20" and 40" with 90 generator sets for electricity sources during transportation. In carrying out the shipping work, the company cooperates with a local feeder vessel operator.

3.2. A generic Indonesian cold chain

First, we analyse the business process in the organizational level between the stakeholders starting from upstream to downstream by using IDEF0 level 0. Then, we draw the business process analysis of the third party logistics company at level 0, which is similar to other Indonesian cold chain providers as shown in Figure 1. The stakeholders are farmers and fisherman located in the Eastern Java, collection centres and production plants located in the Eastern Java, and retailers located in Surabaya and several major cities in the Eastern Java and others located outside Java such as Sulawesi, Kalimantan, and Irian Jaya island. A solid line implies the products flow. Farmers use their own vehicles or take the vehicle directly from the collection centres, while fish and seafood products usually transport using refrigerated trucks or ordinary trucks with Styrofoam boxes containing ice to maintain temperature. The manufacturers produce food products adding value through several preservation and cooling processes to extend the shelf lives. The logistics service provider picks up the products from the collection centres or production plants and deliver to the customers. Third party logistics providers in this study use reefer containers that are transported with the trailer and connected to the diesel in a truck or plug in a vessel as an electrical power source.

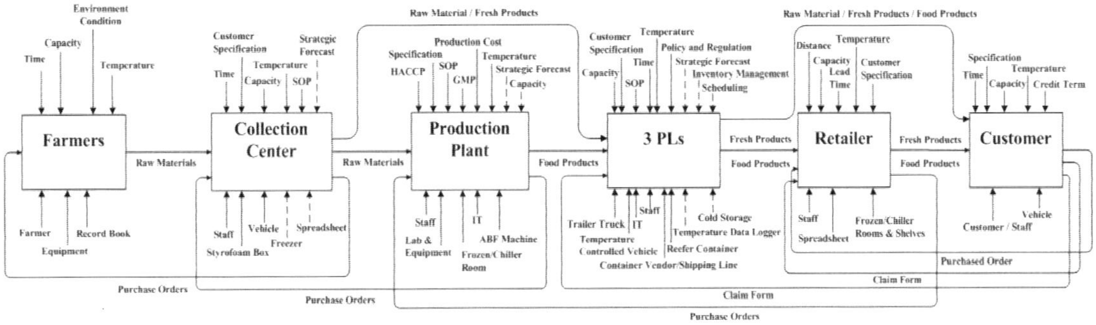

Figure 1. The business process of Indonesian cold chain (IDEF0 Level 0)

3.3. Business process analysis of a third party logistics provider

The business process analysis at level 1 as Figure 2 shows that all activities in third party logistics include plan, source, deliver, and return. From IDEF0 level 1, we analyse the activity in the third party logistics provider in more details and identify problems as observed and interviewed. The marketing department passes orders to the customer service department. Next, the customer service officer checks the availability of containers from the operational department and prepare documents used for shipping the goods. The customer service officer provides container numbers to the operational department for loading and shipping process. Currently, container preparation is based on customer specifications such as container size and container temperature settings according to the product being transported. Currently, the availability of containers is manually checked by the operational staff. Then the container will be carried out to cleaning process and physical check inside and outside the containers. Obviously, the container shortage often occurs in Surabaya container yard because not every container shipping outside Java will be resend. Moreover, in fish harvesting seasons, containers are used to transport fresh fish from outside Java (Eastern Indonesia) to the large island of Java. At present, the company keeps containers outside Java before the fish harvesting season, resulting in containers shortage and lost sales. Therefore, we recommend that the company should implement a strategic forecast using both quantitative and qualitative approaches. Next, the company should perform risk analysis, especially, when there is natural disaster.

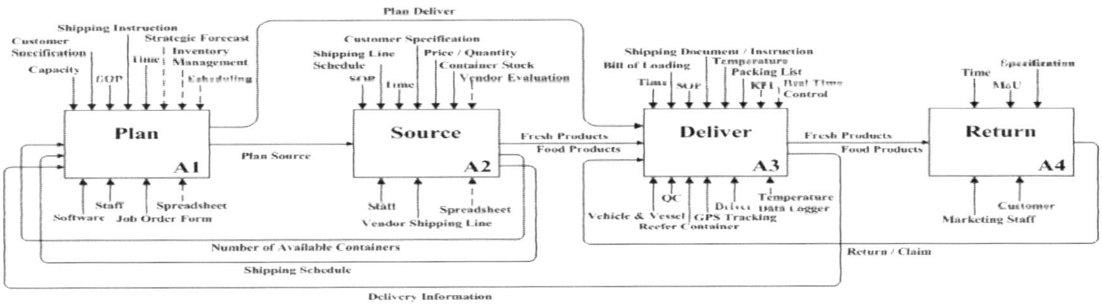

Figure 2. The business process of a third party logistics provider (IDEF0 Level 1)

Customer service officer checks the local carrier operators to get space and plug on the ship before the specified closing time. However, there is often sudden changes in schedule by local carriers. As a result, many shipments are late with additional costs. We recommend the company to evaluate the carrier performance and build the preferred carrier list. The shipment rejection rate is high because the container condition does not comply with the customer specification. In addition, the company does not separate containers used transport raw materials and ready-to-eat food or beverage products. We separate the type of temperature controlled containers used to maintain the quality of the product and to prevent the cross contamination. There is no real time monitoring the temperature, and no coordination between quality control staff and customers. Hence, the company should establish the meeting among the involved staff and invest in the temperature control system. This would cause the higher operating cost, but enhance the food safety and the traceability system which add the value to the customers. The guidelines for improvement at the activity level is shown in Table 1.

Table 1. Summary of current problems and recommendations for improvement

Activities	Problems	Improvement Guideline
Plan	Lack of strategic forecast	Implement the strategic forecast for demand of containers based on type of products and location
	No risk management	Implement risk management
Source	Local carrier is not efficient	Evaluate local carriers and build the preferred carrier list
	Lack of container database	Keep record all data in the spreadsheet or use the system regularly
Deliver	Real time tracking	Real time record and container trucking system
	No coordination between quality control staff and customers	Establish the meeting of multi functional team
	No temperature record	Use temperature data loggers
	No temperature control during loading	Establish temperature controlled system Apply precooling system
	High shipping cancelation from local carriers	Improve the vehicle scheduling and check the local carrier schedule regularly
	High cost of additional plug at a container yard	Establish a cold storage at the right location
Return	High shipment rejection rate	Separate the types of goods in containers or using multi temperature controlled containers

4. Conclusion

The current cold supply chain for food products in Indonesia was explored. We analysed the business process of the third party logistics provider based on IDEF0 level 0 and level 1. Then, we identified the problems occurred in major activities and give the recommendations for improvement such as implementing of the strategic forecast to decrease the container shortage, evaluation the local carrier to reduce the delayed and avoid the additional cost, recoding the data and building the database, monitoring the temperature data to control the temperature, applying the precooling system, and separation of the container for raw materials and finished goods. In this study, we found difficulties in obtaining data, hence, we observed and collected data for a month. In addition, there was limitations in collecting some sensitive information due to company confidential. We hope that our guideline can be implemented at the third party logistics provider to increase the supply chain efficiency. In the future, other third party logistics providers should improve their service management to increase the competitiveness in domestic and prepare to compete with foreign companies.

5. References

[1] Kitinoja L 2013 Use of cold chains for reducing food losses in developing countries *J. Post. Edu.* **13** 03

[2] Gandi I 2016 Mitigating high logistics cost *Proc. Int. Conf. Regional Logistics Provider* (Jakarta, Indonesia, 23 June 2016) p 15

[3] Amran T G 2015 Design of cold chain third-party logistics (3 PL) for halal food in Indonesia *Proc. 8th International Seminar on Industrial Engineering and Management* (Malang, Indoneisa, 17-19 March 2015) pp 49-55

[4] Ongkunaruk P 2015 Business process analysis and improvement for a raw milk collection centre in Thailand *Agriculture and Agricultural Science Procedia* **3** 35-9

[5] Greeff G and Ghoshal R 2004 *Practical E-Manufacturing and Supply Chain Management* (Netherland: Elsevier Book)

[6] Praasertwattanakul Y and Ongkunaruk P 2018 The analysis of a vertically integrated organic rice company: a case study in Thailand *J. Inter. Food* **25** 481-6

Self-Assembly of Aldehyde Lipids in Water

M Janlad[1, 2, 3, 4], P Boonnoy[1, 2, 3, 4] and J Wong-ekkabut[1,2,3,4,*]

[1] Department of Physics, Faculty of Science, Kasetsart University, Bangkok 10900, Thailand.
[2] Computational Biomodelling Laboratory for Agricultural Science and Technology (CBLAST), Faculty of Science, Kasetsart University, Bangkok 10900, Thailand
[3] Thailand Center of Excellence in Physics (ThEP Center), Commission on Higher Education,10400, Thailand.
[4] Specialized Center of Rubber and Polymer Materials for Agriculture and Industry (RPM),Faculty of Science, Kasetsart University, Bangkok 10900, Thailand.

[*] Corresponding author: jirasak.w@ku.ac.th

Abstract. Lipid based self-assembled structures has been applied for drug delivery systems. It is well known that lipid self-assembled structures depend on the chemical structures of lipids, however the self-assembled structures of oxidized lipids with asymmetrical structures remain unclear. In this study, 1-palmitoyl-2-(9-oxo-nonanoyl)-*sn*-glycero-3-phosphocholine (9-al) was used to investigate its self-assembly structure. The molecular dynamics (MD) simulation was performed with the system consisting of 128 lipids and 8,960 simple point charge (SPC) water molecules. Initially, all lipids and water molecules were randomly generated in the simulation box, then molecular trajectories were integrated and collected over microsecond times. The results showed that the equilibrium self-assembled structures of 9-al lipid preferred to form a micelle structure. We analyzed the lipid's angle distribution and the lipid's chain length to determine the lipid geometry. The results are in agreement with the previous MD simulation (Boonnoy, P.; *et al.* J. Phys. Chem. Lett. 2015, 6, 4884– 4888) that showed the deformation of aldehyde lipid bilayer into micelle structure due to their shorter and highly mobile of aldehyde functional group in oxidized chain.

1. Introduction

Oxidized lipids are the products of an oxidative attack on the unsaturated lipids acyl chains by free radicals (OH^-, O^{2-}, etc.)[1]. There are two major oxidized lipids species such as hydroxyl- or hydroperoxy-dieonyl PCs and truncated chains of PCs with aldehyde or carboxylic groups [2]. The previous experimental and computational studies have shown that the presence of oxidized lipid in cell membrane affect the biophysical properties of cell membrane and can lead to pore formation and lipid bilayer disintegration especially in case of the presence of high concentration of oxidized lipid with an aldehyde functional group [3,4]. In the previous study of Boonnoy *et al.* [5] have shown that the presence of 75% and 100% of aldehyde lipid in 1-palmitoyl-2-linoleoyl-sn-glycero-3-phosphatidylcholine (PLPC) lipid bilayer lead to the lipid bilayer deformation into micelle. One possible explanation it that since the geometry of aldehyde lipid is a truncated cone the asymmetrical structure of aldehyde lipid and the extended structure of the oxidized chain. However the self-aggregation of aldehyde lipid remain unresolved. In this work, the self-assembly structure of aldehyde lipid starting from random configuration was investigated by using molecular dynamics simulations. The lipid self-assembly processes and the lipid geometry were analyzed and discussed.

2. Method

Molecular dynamics (MD) simulation of lipid self-assembly were performed using GROMACS 5.1.1 package [6]. 9-al lipid (1-palmitoyl-2-(9-oxo-nonanoyl)-*sn*-glycero-3-phosphocholine), an asymmetrical oxidized lipid structure with an aldehyde functional group in the oxidized chain (see figure 1(A) for the molecular structure) was used to study the self-assembled structures starting from random configuration to equilibrium structures. The united atom force-field parameters of 9-al lipid were taken from previous studies [5,7,8]. The system consists of 128 9-al lipid molecules and 8,960 simple point charge (SPC) water [9]. The initial structure of the system was generated using gmx insert-molecules tool that is provided in the GROMACS package by which all lipid and water molecules were randomly placed in the simulation box as shown in figure 1(B). After energy minimization process using a steeped descent algorithm, MD simulations were run for 1 microseconds with a time step of 2 fs. Periodic boundary condition were applied in all xyz directions and the neighbor list was updated for every time step. The simulation was run under the constant of particle number, pressure and temperature (NPT ensemble). A constant isotropic pressure coupling was controlled using the Parrinello-Rahman algorithm at 1 bar with a time constant of 3.0 ps and the compressibility of 4.5 x 10^{-5} bar^{-1} [10]. A constant temperature at 298 K was controlled using velocity rescaling algorithm [11] with a time constant of 0.1 ps. A cut-off of 1.0 nm was used for the long-range neighbor list of electrostatic and Van der Waals interactions. The Particle-mesh Ewald (PME) method was applied to calculate long-rang electrostatic interactions with the PME grid of 0.12 nm in the reciprocal-apace interactions and cubic interpolation. All bond lengths were constrained using the LINCS algorithm [12].The Visual Molecular Dynamics (VMD) software was used for molecular visualization [13].

3. Results and Discussions

In this section, we discuss the self-assembly of 9-al in water. Figure 1(A) shows the molecular structure of the aldehyde lipids in which an portion of lipid tail contains an oxygen atom. Figure 1(B-D) show the mechanism of 9-al self-assembly in water. The snapshots were taken at simulation times (B) t = 0 ns, (C) t = 214 ns, and (D) t = 1000 ns. The process start with all lipids and water were randomly dispersed in the simulation box. The self-assemble lipids quickly form to small clusters by which the polar lipid head group are in contact with water and the lipid tail tend to stay together. Then the small clusters gradually merge into the large clusters. Finally, the final structure of 9-al lipids are stable to form micelle structure. As the snapshot, we found the 9-al lipids bend the polar lipid tails in contact with water, therefore the average geometries of 9al-lipids is conical leading to form micelle structure.

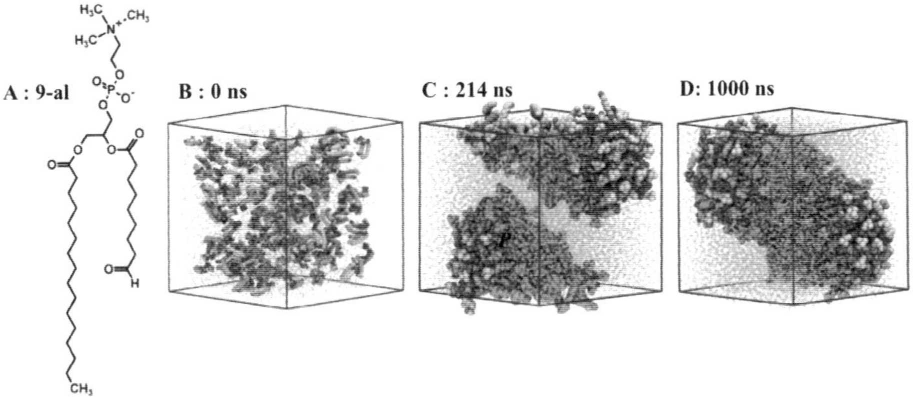

Figure 1. (A) Molecular structure of 9-al lipid (1-palmitoyl-2-(9-oxo-nonanoyl)-*sn*-glycero-3-phosphocholine). (B-D) The illustrations of self-aggregation of 9-al lipids in water. (B) The initial configuration of a simulated system. (C) Lipid gradually incorporates by bend the polar head contact with water. (D) 9-al lipids form the micelle structure. 9-al lipids are shown in green spheres, with phosphorus atom as purple spheres. Oxygen atoms are shown in red. Water is shown in light blue region.

Based on the final structure of lipid self-assembly, the shape of individual lipids were analyzed to explain how lipids prefer to form those structures. First, the distances between the phosphate atom in lipid headgroup and the last atom in the *sn-1* and *sn-2* chains (as defined l_1 and l_2, respectively) were calculated to characterized the tail lengths. In addition, the separation between the *sn-1* and *sn-2* chains was characterized by the distance between the last atoms in lipid tails, l_3. The results in Table 1 show that the l_1 , l_2 and l_3. The ratio l_1:l_2 is greater than 1 due to the truncated lipids in the *sn-2* and the ratio l_2:l_3 is 0.65. The previous work of Boonnoy *et al.* [5] also showed that the ratio l_2:l_3 of the aldehyde lipids were significantly smaller than 1. Finally, the interior angles θ_A, θ_C were analyzed to determine the lipid geometry. The interior angles θ_A is averaged the angle between vectors of the head group to the last atom in the *sn-1* and *sn-2* chains. The interior angles θ_C is the angle between l_1 and l_3. As the figure 2 shown the angle distribution of θ_A and θ_C for 9-al lipids. The $\langle\theta_A\rangle$ and $\langle\theta_C\rangle$ of 9-al are were found at 74±2° and 37±1°, respectively. These results were in agreement with the previous study [5].

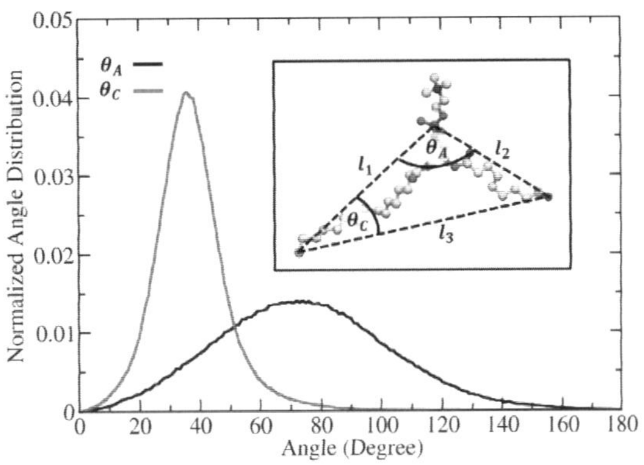

Figure 2. Angle distribution of θ_A and θ_C for 9-al cylindrical micelle. The definition of angle θ_A , θ_C , l_1, l_2 and l_3 are shown in the inset.

Table 1. The average angle θ_A , θ_C and the lengths of lipid chain ($l_1, l_2, and\ l_3$). See also figure 2.

	θ_A (degree)	θ_C (degree)	l_1 (nm)	l_2 (nm)	l_3 (nm)	$l_1{:}l_2$	$l_2{:}l_3$
9-al	74±2	37±1	1.91±0.01	1.27±0.01	1.95±0.03	1.50	0.65
9-al[a]	74±2	37±1	1.86±0.16	1.26±0.01	1.93±0.04	1.48	0.65

[a]Data were taken from a previous study by Boonnoy *et al.* [5]

Definition of angle and length. The interior angles of aldehyde lipid tails are represented by θ_A , θ_C . l_1 and l_2 are represent the lengths of the *sn-1* and *sn-2* lipid tails, respectively, and l_3 is the distance between the end of lipid tails.

4. Conclusion

MD simulation of aldehyde lipid self-assembly were performed to ensure that the lipid geometry play an essential role in the self-aggregation structure. The results showed that 9-al lipid spontaneously form stable cylindrical micelle structure similarly to the previous computer simulation [5]. The analysis of structural properties are in good agreement for all different initial configurations. Our results confirmed that the presence of aldehyde lipid can strongly modify membrane structure. An asymmetrical structures and the extended of aldehyde functional group in the oxidized chain can lead to form the micelle structure.

5. References

[1] Brodnitz M. H, Nawar W. W, Fagerson I. S 1968 *Lipids.* **3** 59-64
[2] Jurkiewicz P, Olżyńska A, Cwiklik L, Conte E, Jungwirth P, Megli F. M, Hof M 2012 *Biochim. Biophys. Acta, Biomembr.* **1818** 2388-402
[3] Cwiklik L, Jungwirth P 2010 *Chem. Phys. Lett.***486** 99-103
[4] Weber G, Charitat T, Baptista M. S, Uchoa A. F, Pavani C, Junqueira H. C, Guo Y. C, Baulin V. A, Itri R, Marques C. M, Schroder A. P 2014 *Soft Matter.* **10** 4241-7
[5] Boonnoy P, Jarerattanachat V, Karttunen M, Wong-ekkabut J 2015 *J. Phys. Chem. Lett.* **6** 4884-8
[6] Abraham M. J, Murtola T, Schulz R, Páll S, Smith J. C, Hess B, Lindahlad E 2015 *SoftwareX.* **1** 19-25
[7] Bachar M, Brunelle P, Tieleman D. P, Rauk A 2004 *J. Phys. Chem. B* **108** 7170-9
[8] Wong-Ekkabut J, Xu Z, Triampo W, Tang I-M, Tieleman D. P, Monticelli L. 2007 *Biophys. J.* **93** 4225-36
[9] Berendsen H. J, Postma J. P, van Gunsteren W. F, Hermans J 1981 *Springer* 331-42
[10] Parrinello M, Rahman A 1981 *J. Appl. Phys.* **52** 7182-90
[11] Bussi G, Donadio D, Parrinello M 2007 *J. Chem. Phys.* **126**
[12] Hess B, Bekker H, Berendsen H. J, Fraaije J. G 1997 *J. Comput. Chem.* **18** 1463-72
[13] Humphrey W, Dalke A, Schulten K 1996 *J. Mol. Graph.* **14** 33-8

Acknowledgements

This work was financially supported by Kasetsart University Research and Development Institute (KURDI) and Faculty of Science at Kasetsart University (JW). MJ acknowledges the Institute for the Promotion of Teaching Science and Technology (IPST) and the Development and Promotion of Science and Technology Talents Project (DPST). The support of the Thailand Research Fund (TRF) through the Royal Golden Jubilee Ph.D. Program (Grant No.PHD/0204/2559) and the TRF Research Scholar Program (Grant No. RSA6180021) for PB and JW, respectively is acknowledged.

IOP Conf. Series: Materials Science and Engineering **526** (2019) 012006 doi:10.1088/1757-899X/526/1/012006

Monte Carlo simulations of nanorod filler in composite polymer material

N Kerdkaen[1,2,3], T Sutthibutpong*[2,3,4], S Phongphanphanee[2,5,6], S Boonchui[1,6], and J Wong-ekkabut*[1,2,3,6]

[1] Department of Physics, Faculty of Science, Kasetsart University, Bangkok 10900, Thailand.
[2] Computational Biomodelling Laboratory for Agricultural Science and Technology (CBLAST), Faculty of Science, Kasetsart University, Bangkok 10900, Thailand
[3] Thailand Center of Excellence in Physics (ThEP Center), Commission on Higher Education, Bangkok 10400, Thailand
[4] Department of Physics, Faculty of Science, King Mongkut's University of Technology Thonburi (KMUTT), Bangkok 10140, Thailand
[5] Department of Material Science, Faculty of Science, Kasetsart University, Bangkok 10900, Thailand
[6] Specialized Center of Rubber and Polymer Materials for Agriculture and Industry (RPM), Faculty of Science, Kasetsart University, Bangkok 10900, Thailand

* Corresponding authors: jirasak.w@ku.ac.th , thana.sut@mail.kmutt.ac.th

Abstract Conductive polymer (CP) is a special class of polymeric materials with conductive property which possesses high potential for the fabrication of future advanced electronic devices. To design a higher conductivity polymer, composite polymer material with conductive nanorod filler is one of alternative ideas to replace the high-cost intrinsically conductive polymeric materials. In this study, our in-house Monte-Carlo simulation was performed to generate a number of randomly-positioned nanorods within a 3D confined-space box and to assess percolation paths formed by nanorods that connected two electrodes for each configuration. The results showed that the probability of finding connection path was related to the concentrations of nanorod filler in 3D confined-space box. The increasing probability to find connection paths when adding more filler concentrations was found to be a logistic growth, in which growth rate and threshold concentration depended on soft-shell filler radius. Our finding will be beneficial for designing composite conductive polymer for switching sensor.

1. Introduction

Conductive polymer (CP) is a special class of polymeric materials with electronic and ionic conductivity [1] that can be used in many applications such as sensors and rechargeable battery. Moreover, their organic properties can be used to control drug delivery system. However, CPs are expensive and poorer mechanical properties than other classes of polymeric materials [1]. One of the solutions to develop the properties of conductive polymer is to design the composite polymeric materials with conductive nanoparticles filler. The conductive polymers nanocomposites [2] can be used as an alternative for CPs. In order to obtain the design for nanocomposites with desired mechanical properties, computer simulations have been used as a virtual experiment to observe behavior of system that greatly reduce time consuming and cost from trial and error processes in the laboratories. However, the complex system, such as nanocomposites, still consume remarkably large amount of computing resources. Therefore, the system must be simplified to reduce computational time. In this study, a polymer composite system with nanowire fillers was represented by a closed cubic box filled with uniformly sized spherocylinder. The electrical percolation behaviors at different concentrations, electronic interaction ranges and sizes of the confined-space box were investigated, as the composite became conductive when the filler concentration reached the percolation threshold [3]. The effects of spherocylinder filler concentrations on the conductivity were determined from the probability of percolated path formation in a large number of nanorod configurations.

2. Methodology

2.1. Hard-core with Soft-shell Spherocylinder Model for Nanorod Fillers

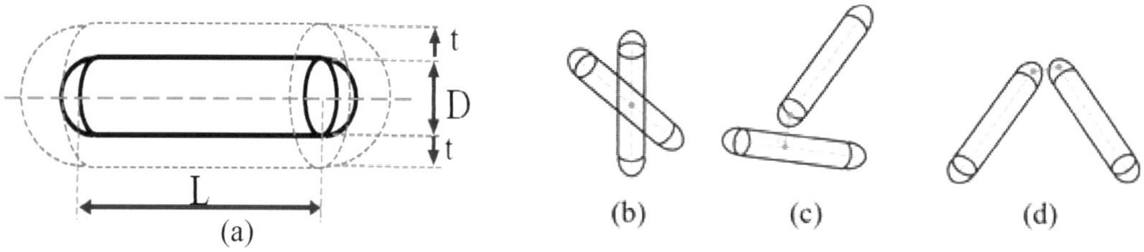

Figure 1. (a) A filler particle model consisting a spherocylinder hard-core made of two hemispheres of diameter (D) connected by a cylinder of length (L) with similar diameter and the outer of thickness (t). Three possible contacts patterns for nanorod electrical consist of (b) side-to-side, (c) side-to-end and (d) end-to-end.

In this study, the rod-like filler particles were represented by 3D spherocylinder-shaped particles [4]. Each of them consists of a rigid and impenetrable inner hard-core to represent the nanorod material and the penetrable outer soft-shell to represent the effective range of electron tunneling effect and the interactions between the matrix and filler (see figure 1a). If the soft-shell of a rod penetrate to touch the hard-core of another rod (The minimum distance between filler was less than sum of a soft-shell filler thickness and twice a hard-core filler radius.), they are considered to be in electrical contact Figure 1(b-d) display three possible types of electrical connections in the model [4].

2.2. Software Methods and Simulation Setup

Our in-house python program was developed to generate a number of randomly-positioned individual spherocylinder filler units in a confine-spaced box (Nanorods were only allowed to be within the box.) and to determine the probability to find a percolated path that represent the conduction of the composite. The program consisted of three main parts. In the first part, the program randomize the position and orientation for each filler unit within the confine-spaced box and avoid overlapping between the hard-cores of all fillers. Then, in the second part, the program determined the list of connected filler units as defined in section 2.1 and searched for percolated path. The last part of the program considered whether existence of a percolated path. A continuous network of nanorod filler units formed between two opposite sides of the confined-space box in the system is defined as 'conductivity' (see figure 2). The processes were repeated many times in order to calculate the probability of conductivity for each system condition.

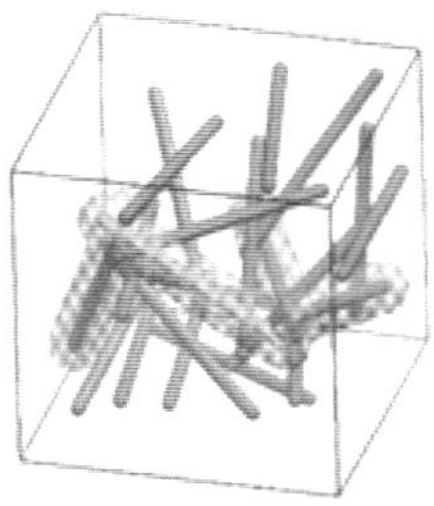

Figure 2. Example of confine-spaced box. The percolated or 'conductive' path is shown in cyan.

In this study, dimensions of the nanorods and the confined-space box, nanorod concentrations and electronic interaction ranges were varied to see their effects on conductivity. Hard-core part of a spherocylinder filler length is 20 Arbitrary Unit (AU) and diameter is 2 AU. Filler concentration was varied from 0 to 15 % by volume with 0.5 intervals. Confined-space box was a cube of length dimension 30, 50, 70 and 100 AU. Soft-shell thickness (t) was varied from 0.8 to 3.2 AU with 0.4 interval to observed the impact of the electronic interaction ranges on the electrical conductivity.

3. Results and discussions

3.1. Probability of Electrical Conductivity

In this study, our in-house python program be used to simulation confined-space box matrix with uniformly sized spherocylinder filler model to analyze the probability of electrical conductivity(P) as a function of confined-space box sizes, filler's concentrations(C) and filler soft-shell thicknesses. The results for several conditions that varied filler's concentrations and soft-shell thicknesses (constant confined-space box size) are shown in figure 3.

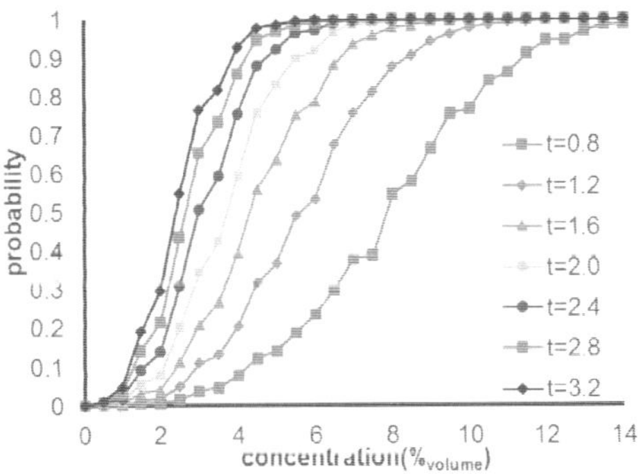

(a)

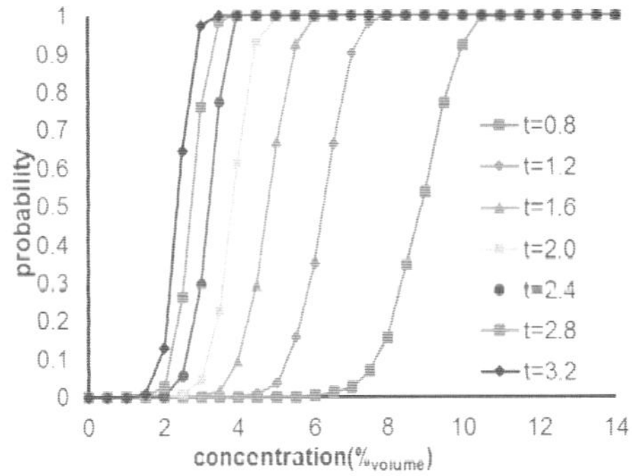

(b)

Figure 3. Percolation probability of nanorod filler confined within the confine-spaced cube of dimensions (a) 30 AU and (b) 100 AU. 3000 samples were calculated as function of nanorod filler concentration at different soft-shell inner thickness (t). The nanorod filler length (L) was set at 20 AU.

As figure 3 shown, the conductivity probability increases slowly from zero when filler's concentration rise up, and the increasing positions from zero depend on soft-shell thickness. When concentration rise up, the filler's concentration at the half of maximum probability is called "critical filler's concentration". Critical filler's concentration depends on soft-shell thickness and confined-space box size. At high filler's concentration, probability of electrical conductivity at any soft-shell thickness rise up to maximum probability. Each curve line in figure 3 is concordant to basic shape of the curve of a logistic function. Therefore, relationship of the systems can be described by logistic function.

3.2 Logistic Function

Logistic function is a common "S" shape curve with equation (1):

$$f(x) = \frac{L}{1+e^{-K(X-X_0)}}. \quad (1)$$

Where X_0 is the x-value of curve's midpoint, L is the curve's maximum value, and K is the steepness of the curve. Variable in equation (1) can be compared to variable of logistic function. Critical filler's concentration (C_0), maximum probability and growing rate of probability (α) are standed for X_0, L and K respectively. Therefore, the applicable function for this study is shown in equation (2),

$$P(C) = \frac{1}{1+e^{-\alpha(c-c_0)}} \quad (2)$$

Logistic function variable that fitting curve to the result be found as shown in figure 4. The fitting parameters α and C_0 are shown in figure 4 (a) and 4 (b), respectively. Both figures show that α and C_0 depended on soft-shell filler thickness in any confined-space box size but the ratio are different. The results suggested that probability of electrical conductivity not only depends on filler's concentration but also on soft-shell filler thickness.

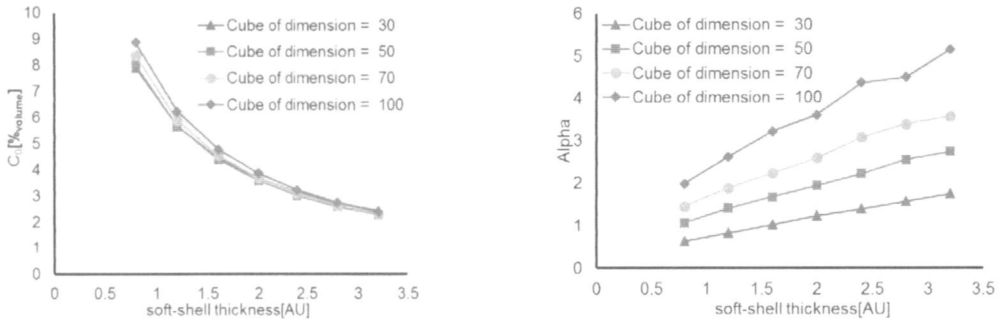

(a) (b)

Figure 4. (a) critical filler concentrations (L_0) as function of soft-shell thickness (t) at different box dimension and (b) probability growth rate (α) as function of soft-shell thickness (t) at different box dimension

4. Conclusions

The in-house python program is capable of generating a number spherocylinder filler in a confined-space box and adjusting several parameters such as filler concentrations and soft-shell radii in order to calculated probability of electrical conductivity on the basis of Monte Carlo simulations. The results show that probability of electrical conductivity is related to logistic function of filler concentration. The critical filler concentrations (C_0) and probability growth rate (α) are depended on soft shell thickness (t). This program can be used as an aid in studying and fabrication conductive composite polymer in switching stress sensor.

5. References

[1] Ravichandran R, Sundarrajan S, Venugopal JR, Mukherjee S and Ramakrishna S 2010 *J R Soc Interface* **7** 559-579
[2] Gangopadhyay, R, and De, A 2000 *Chem. Mater* **12** 608-622
[3] Kalyon, D M, Birinci, E, Yazici, R, Karuv, B, and Walsh, S 1999 *J Mater Sci* **32** 5551-5560
[4] Belashi A 2011 *Percolation modeling in polymer nanocomposites* (Toledo: University of Toledo) p 525

Acknowledgements

This work was financially supported Thailand Research Fund (TRF) (Grant No. RDG60T0148).

Bond strength of laminated films between poly(lactic acid) based cast films and metalized cast polypropylene films

N Hongsriphan[1], J Chaichana, C Pornpitakdamrong, and S Thaptham

[1] Department of Materials Science and Engineering, Faculty of Engineering and Industrial Technology, Silpakorn University, Sanam Chan Palace Campus, Nakhon Pathom 73000, Thailand .

* Corresponding author: Hongsriphan_N@su.ac.th

Abstract. This research was carried out to study the feasibility to use biodegradable poly(lactic acid) (PLA) in flexible packaging films, which they were usually laminated with metalized cast polypropylene (MCPP) film in order to reduce oxygen permeability and could be microwaveable. Since PLA was too brittle, two types of polymers; poly(butylene adipate-co-terephthalate) (PBAT) and polyethylene glycol (PEG), were blended and extruded into cast films in the weight ratios of 5, 10, and 15 wt%. The prepared blend films were laminated with MCPP films using a commercial polyurethane adhesive. Mechanical and thermal properties of the blend films were investigated by tensile testing and differential scanning calorimetry. The bond strength of the laminated films was analyzed. It was found that the glass transition temperatures of PLA in the PLA/PEG blends were dramatically reduced attributed to the plasticizing effect, however, they were not changed in the PLA/PBAT blends. As a result, tensile strength of the PLA/PEG films was reduced significantly with respect to PEG content. The laminated films of PLA/PBAT blend and MCPP had higher bond strength than those of PLA and MCPP or PLA/PEG and MCPP. The transparent PLA/PBAT 95/5 wt% was the best formula that could replace poly(ethylene terephthalate) for laminated film.

1. Introduction

Typically, flexible packaging is produced from polyethylene (PE), polypropylene (PP), and poly (ethylene terephthalate) (PET). However, these flexible packaging films require to use as laminated films for specific end-use, such as oxygen barrier and light barrier for long-time shelf life. Metalized cast polypropylene (MCPP) film [1] is usually laminated with other polymer substrates for this purpose with lower cost. Also, the laminated film with MCPP can be microwaveable. These polymers are non-biodegradable, so that they will be accumulated as plastic waste to cause environment problem. Use of biodegradable polymer such as poly(lactic acid) [2] has been encouraged to use. Since PLA was too brittle, two types of polymers; poly(butylene adipate-co-terephthalate) (PBAT) and polyethylene glycol (PEG), was melt blended into PLA to produce cast films. The lamination was performed by using a commercial polyurethane adhesive to bind the blend films with the MCPP films. Therefore, the bond strength or adhesive strength [3] was evaluated and compared with the use of pure PLA film.

2. Experimental

2.1 Materials

Poly(lactic acid) or PLA (Ingeo™ Biopolymer 4043D, density of 1.24 g/cm^3, MFI of 6 g/10 min (210°C/2.16 kg)) was purchased from NatureWork LLC, USA. Poly(butylene adipate-co-terephthalate) or PBAT (Ecoflex® F BX 7011, a density of 1.25 g/cm^3, MFI of 2.7-4.9 g/10 min (190°C/2.16 kg)) was purchased from BASF, Germany. Polyethylene glycol or PEG (M_n of 5000-7000, density of 1.13 g/cm^3) was purchased from Sigma-Aldrich, USA. Polyurethane adhesive was a mixture between Loctite Liofol LA 7728 (isocyanate part) and Loctite Liofol LA 6028 (hydroxyl part) in the weight ratio of 100/80. They were kindly supported by Henkel (Thailand) Company, Thailand. Metalized cast polypropylene (MCPP) with thickness of 25 μm was kindly provided by Huhtamaki (Thailand) Company, Thailand.

2.2 Melt blending and cast film extrusion

PLA and PBAT pellets were dried in an air-circulating over (Model FD53, Binder, Germany) at 60°C for 8 hours. The melt blending was carried out in a co-rotating twin-screw extruder (SHJ-25, Yongteng, China) with the weight ratios of the PLA/PBAT or PLA/PEG blend were 100/0, 95/5, 90/10, and 85/15 wt%. The barrel/die temperature was 140-190°C, and the screw-speed was 40 rpm. The extrudate was cooled in a water bath and pelletized. After thoroughly drying, the pure PLA and the blend pellets were fabricated into films using a cast film extruder (LBCR-150, Labtech Engineering Co, Thailand). The barrel/die temperature for cast film extruder was 180-210°C. The chill-roll temperature was set at 28°C. The chill-roll speed was 1.1-1.3 m/min, and the pull-out speed was 2.2-2.6 m/min. For each blend ratio, the condition was justified to obtain the cast film that had the thickness between 60 to 90 μm.

2.3 Lamination between cast films and MCPP films

Loctite Liofol LA 7728 (poly-isocyanates) and Loctite Liofol LA 6028 (polyols) in the weight ratio of 100/80 were mixed and warmed in a water bath at 40°C. Then, the adhesive was poured onto the substrate surface (prepared cast film), and the Meyer rod was rolled over manually with constant speed to obtain the adhesive thickness of about 5 μm. After that, the MCPP film was placed on top of the adhesive layer. A rubber roller was rolled over the laminated films several times to ensure adhesion of the laminated films.

2.4 Tensile testing of cast films

The tensile testing of blend films was carried out in accordance to ASTM-D882 using a universal testing machine (Instron 5969, Instron Engineering Corporation, USA). The specimens were cut from cast films by a cutter in the machine direction to obtain the rectangular strip of 2.54 x 15.24 cm. The gauge length was 100 mm. The load cell was 5 kN. The crosshead speed was 10 mm/min. Five trials were tested, which the averages and standard deviations were reported graphically.

2.5 Thermal analysis by differential scanning calorimetry (DSC) of blend films

The pure PLA and the blend films were cut into small pieces (5-7 mg) for thermal property analysis by a differential scanning calorimeter (DSC Pyris 1, Perkin Elmer, USA). The tests were performed in a heat-cool-reheat mode from 25 to 250°C with a heating/cooling rate of 5°C/min. The glass transition temperature, the cold crystallization temperature, and the degree of crystallinity were reported.

2.6 Bond strength testing of laminated films

The bond strength was evaluated in accordance to ASTM-F904 (comparison of bond strength or ply adhesion of similar laminates made from flexible materials). The specimen was cut from laminated films in the dimension of 2.54 x 15.24 cm. The laminated films were set to perform the 180° peel test, which a gauge length was set at 50 mm. The peel speed was set at 300 mm/min. The average load required to separate the laminated was recorded and expressed as the bond strength in Newton (N) per 25 mm.

3. Results and discussion

3.1 Tensile properties of blend films

In this study, the prepared blend films were transparent but they were less transparent than the pure PLA film. Table 1 shows film thickness of the pure PLA film, the blend films, and the laminated films. The thickness of MCPP film was constant at 25.6 ± 1.9 μm. The thickness of the PLA/PEG 85/15 wt% blend films was significant thicker than other films. Since this blend ratio had the lowest melt strength during the cast film extrusion, the chill-roll speed had to be reduced to obtain the continuous films.

Table 1. Film thickness of pure PLA film, the blend films, and the laminated films.

	PLA100	PLA95/ PBAT5	PLA/90/ PBAT10	PLA85/ PBAT15	PLA95/ PEG5	PLA90/ PEG10	PLA85/ PEG15
Cast film (μm)	86.0±0.013	63.0±0.003	71.0±0.005	77.0±0.004	90.0±0.007	81.0±0.008	119.0±0.011
Adhesive layer (μm)	9.0+0.007	7.0±0.005	8.0±0.006	8.0±0.009	6.0±0.005	8.0±0.004	7.0±0.004
Laminated film (μm)	120.0±0.01	95.0±0.005	104.0±0.005	110.0±0.006	121.0±0.002	114.0±0.005	151.0±0.011

Figure 1 presents tensile modulus and tensile strength of the PLA/PBAT blend and the PLA/PEG blend films in the blend weight ratios of 0, 5, 10, and 15 wt%. The pure PLA films had tensile modulus of 4160 ± 363 MPa, and tensile strength of 37.0 ± 1.9 MPa. Since PBAT was more flexible than PLA, blending it into PLA reduced rigidity of the blend films with respect to PBAT content. The tensile modulus of the PLA/PBAT blend films were 3788 ± 342, 3575 ± 379, and 3385 ± 310 MPa for the blend weight ratio of 5, 10, and 15 wt%, respectively. For blending with PEG, the tensile modulus of the blend films were 3357 ± 369, 3457 ± 358, and 902 ± 96 MPa for the blend ratios of 5, 10, and 15 wt%, respectively. Although the blend film with PEG of 15 wt% was the thickest film (due to the lowest melt strength), the tensile modulus was the lowest attributed to the plasticizing effect of PEG molecules. The reduction of tensile strength after blending PLA with PBAT or PEG was in the same fashion.

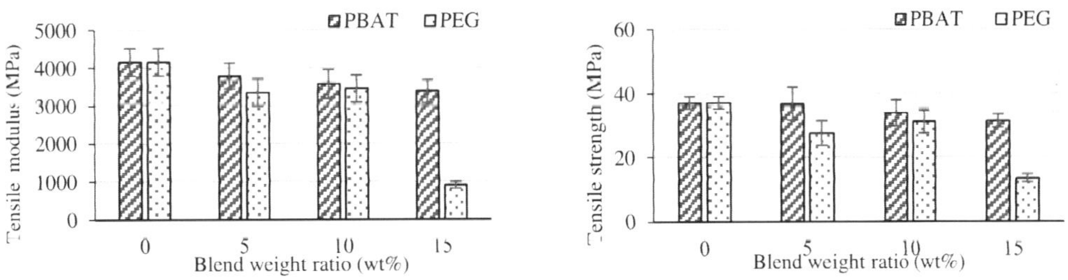

Figure 1. Tensile modulus and tensile strength of the PLA/PBAT and PLA/PEG blend films in the blend weight ratios of 0, 5, 10, and 15 wt%.

3.2 Thermal properties by differential scanning calorimetry

Figure 2 presents the glass transition temperature (T_g), the cold crystallization temperature (T_{cc}), and the degree of crystallinity (X_c) of the PLA/PBAT and PLA/PEG blend films in the blend ratios of 0, 5, 10, and 15 wt%. The blending PLA with PBAT did not change either T_g or T_{cc} and the X_c was increased slightly. In contrast, blending PLA with PEG reduced T_g significantly confirming the plasticizing effect of PEG molecules [4] that interacted with PLA molecules causing more free volume for easier molecular mobility. This influence reduced tensile modulus and tensile strength of the PLA/PEG blend films significantly. In addition, the mobile PLA molecules could recrystallize during the heating scan resulting to dramatically reduction of T_{cc}. As a result, the X_c of PLA was increased from 8% to 29% when blending with PEG of 15 wt%. Despite of higher X_c, all blend films were still transparent.

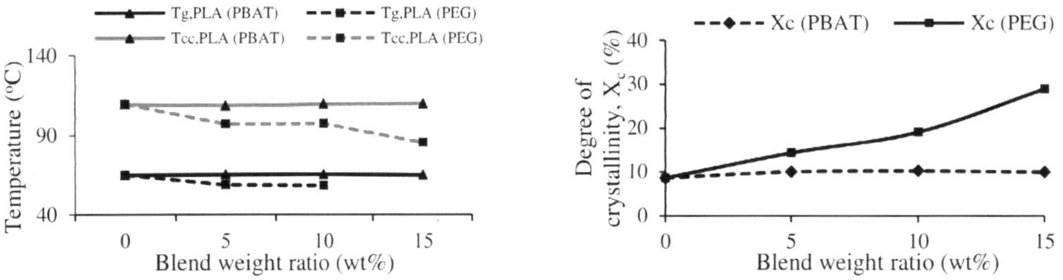

Figure 2. Glass transition temperatures (T_g), cold crystallization temperatures (T_{cc}), and the degree of crystallinity (X_c) of the PLA-based blend films in the blend weight ratios of 0, 5, 10, and 15 wt%.

3.3 Bond strength of laminated films

Figure 3 shows failure of the laminated films after the 180° peel testing. There was trace of aluminium on the cast film surface indicating that aluminium coating layer was peeled out from the MCPP film. The cast films were intact during the delamination process, while the relatively thin MCPP films were stretched and worn. Since the adhesive mixture consisted of poly-isocyanates, it could interact with the hydroxyl groups of the polyester cast films creating high bonding strength [5]. The laminated films of PLA/PEG blend and MCPP showed mostly substrate failure that the deposited aluminium layer was ripped off from the MCPP film. This implied that the bond strength was similar in all blending formula. On the other hand, the laminated films of PLA/PBAT blend and MCPP showed a mix of substrate failure (deposited aluminium layer from the MCPP film) and adhesion failure. The maximum bond strength was obtained in the laminated film between PLA/PBAT 95/5 wt% cast film and MCPP film.

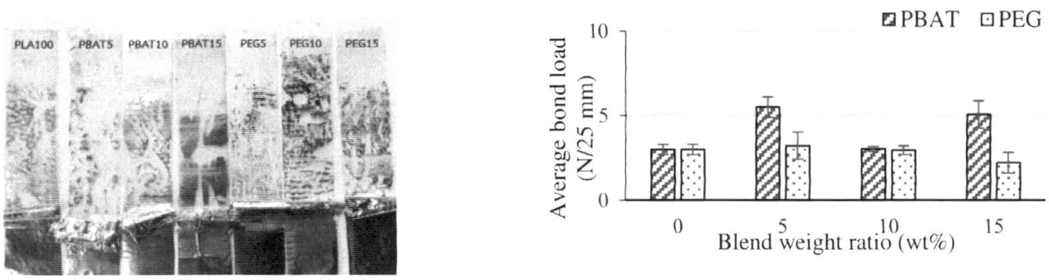

Figure 3. Images of film specimen after failure from the 180° peel test (left), and bond strength of the laminated films between PLA/PBAT or PLA/PEG and MCPP films (right).

4. Conclusion

The laminated films were successful prepared between the PLA/PBAT or the PLA/PEG cast films and the MCPP films. By using the proper polyurethane adhesive, it was found that the transparent PLA/PBAT 95/5 wt% was the best formula that could replace PET for flexible packaging film.

5. References

[1] Gupta S, Dixit M, Sharma K and Saxena NS *Surf Coat Technol* 2009 **204**(5) 661
[2] Hamad K, Kaseem M, Ayyoob M, Joo J and Deri F *Prog Polym Sci* 2018 **85** 83
[3] Jesdinszki M, Struller C, Rodler N, Blondin D, Cassio V, Kucukpinar E and Langowski H-C *J Adhes Sci Technol* 2012 **26**(20-21) 2357
[4] Thongpin C, Tippuwanan C, Buaksuntear K and Chuawittayawuta T *Key Eng Mater* 2017 **751** 337
[5] Sheikhy H, Shahidzadeh M, Ramezanzadeh B and Noroozi F *J Ind Eng Chem* 2013 **19**(6) 1949

IOP Conf. Series: Materials Science and Engineering **526** (2019) 012008 doi:10.1088/1757-899X/526/1/012008

Effect of Acetic Acid Pre-Treatment on the Growth Temperature of Graphene on Copper by Thermal Chemical Vapor Deposition

P Kokmat[1], N Donnuea[2], N Nuntawong[3], AWisitsoraat[3] and A Ruammaitree [1,*]

[1] Department of Physics, Faculty of Science and Technology,
Thammasat University, Pathumthani, 12120, Thailand
[2] Department of Gems and Jewelry, Faculty of Gems,
Burapha University, Chanthaburi IT Campus, Chanthaburi 22170, Thailand
[3] Carbon-based Devices and Nanoelectronics Laboratory, National Electronics and Computer Technology Center, National Science and Technology Development Agency, Pathumthani, 12120, Thailand

* Corresponding author: u4605070@hotmail.com

Abstract. Graphene is a flat monolayer of carbon atoms arranged in a honeycomb lattice. Graphene has many outstanding properties such as high electron mobility, superb strength, transparency and great flexibility. These exotic properties make graphene a promising candidate for solar cell, fuel cell, battery, supercapacitor and so on. In this report, we have studied the effect of acetic acid pre-treatment on the growth temperature of graphene on copper by thermal chemical vapor deposition (CVD). In case of graphene growth without acetic acid pre-treatment, the high temperature of 1000 °C is needed for the growth of graphene by CVD. In contrast, the growth temperature can be decreased to 800 °C for graphene growth with acetic acid pre-treatment. We found that acetic acid pre-treatment can eliminate copper oxide from the sample surface resulting in the growth temperature of graphene decreases.

1. Introduction

Graphene is a flat monolayer of carbon atoms arranged in a honeycomb lattice. Graphene has many outstanding properties such as high electron mobility [1], superb strength [2], transparency [3] and great flexibility [4]. These exotic properties make graphene a promising candidate for solar cell [5], fuel cell [6], battery [7], supercapacitor [8] and so on. Graphene can be fabricated in many methods such as mechanical exfoliation of graphite [9], annealing SiC substrates [10-13] and chemical vapor deposition (CVD) of hydrocarbon gas on metal surface [14-15]. The growth of graphene on copper by CVD has been attracted attention since the high-quality monolayer graphene can be grown on copper due to its low carbon solubility [16]. However, the growth of high-quality and large-area monolayer graphene is still challenging because the conditions of CVD and copper surface have to be controlled intensively.

Pre-treatment of copper surface is a necessary process for the growth of graphene on commercial copper by CVD because it can remove the impurity and copper oxide from the copper surface.

IOP Conf. Series: Materials Science and Engineering **526** (2019) 012008 doi:10.1088/1757-899X/526/1/012008

Chemical etching is a popular procedure for etching of copper surface [17]. Nevertheless it cannot treat the extreme-thin copper because the chemical etching reduces the thickness of copper. Boiling copper in acetic acid is a pre-treatment for eliminating oxide layer from the copper surface without etching. Therefore, the acetic acid pre-treatment becomes a suitable method for cleaning the copper surface. In this report, the effect of acetic acid pre-treatment on the growth of graphene on copper by CVD is studied. The presence of copper oxide is confirmed using Raman spectroscopy and energy dispersive X-ray spectroscopy (EDX).

2. Experiment

Copper plates with the thickness of 0.18 mm were cut to dimension of 40 mm×30 mm followed by boiling in acetic acid at 100 °C for 20 minutes. After that, the copper plates were cleaned by ultrasonic cleaner in acetone for 5 minutes. After the acetone evaporated, a copper plate was put in a quartz tube of CVD furnace. The air was evacuated from the quartz tube until the base pressure reached ~10^{-3} torr and then heated to 800-1000 °C with an average heating rate of 15 °C per minute. At the growth temperature, acetylene gas with flow rate of 200 sccm was introduced into the quartz tube. The copper plate was annealed for 30 minutes under a pressure of 1 torr. After that the acetylene gas was stopped and the sample was cooled down rapidly to room temperature.

The presence of graphene and copper oxide on the copper plate was confirmed by Raman measurements carried out at room temperature using a Renishaw spectrometer with a 100x objective and a 532 nm laser. The laser beam size is 1 μm in diameter. Surface morphology and element analysis were measured using field-emission scanning electron microscope (FE-SEM) equipped with an energy dispersive analysis of X-rays (EDX) facility (Jeol) at incident beam of 15 keV.

3. Results and discussion

Figure 1 shows Raman spectra of the samples annealed at 800-1000 °C with/without acetic pre-treatment. In the case of samples annealed at 800 °C and 900 °C without acetic pre-treatment, the Raman peaks at 280, 330 and 612 cm^{-1} are clearly appeared. The presence of these Raman peaks reveals that the surfaces of these samples are covered by CuO [18]. The absence of G peak (~1584 cm^{-1}) and 2D (~2700 cm^{-1}) confirms that there is no graphene on these samples. For the sample annealed at 1000 °C without acetic pre-treatment and the sample annealed at 800 °C and 900 °C with acetic pre-treatment, the Raman spectra show D, G and 2D peaks at ~1350 cm^{-1}, ~1584 cm^{-1} and ~2700 cm^{-1}, respectively. The presence of G and 2D peaks of these samples confirms the presence of graphene on the sample surfaces. Figure 2 shows SEM images of the sample annealed at 800 °C and 900 °C with/without acetic pre-treatment. The presence of oxide on the surfaces is also confirmed by EDX as shown in Table 1. However, the Raman spectroscopy and EDX are measured at different position therefore the appearance of little weight% of oxygen in the sample annealed at 800 °C with acetic pre-treatment is possible.

The results of Raman spectroscopy and EDX reveal that the presence of CuO on the sample surface obstructs the formation of graphene therefore the elimination of CuO before the growth of graphene on Cu is needed. The sample annealed at 1000 °C shows that the CuO can be removed from the surface at annealing temperature of 1000 °C. In addition, the acetic pre-treatment can eliminate CuO by reacting with CuO to form cupric acetate by the following reactions [19]

$$CuO + 2CH_3COOH \rightarrow Cu(CH_3COO)_2 + H_2O$$

The cupric acetate can be eliminated from the surface at low temperature (~240 °C) therefore the acetic acid pre-treatment can lower the growth temperature of graphene on copper.

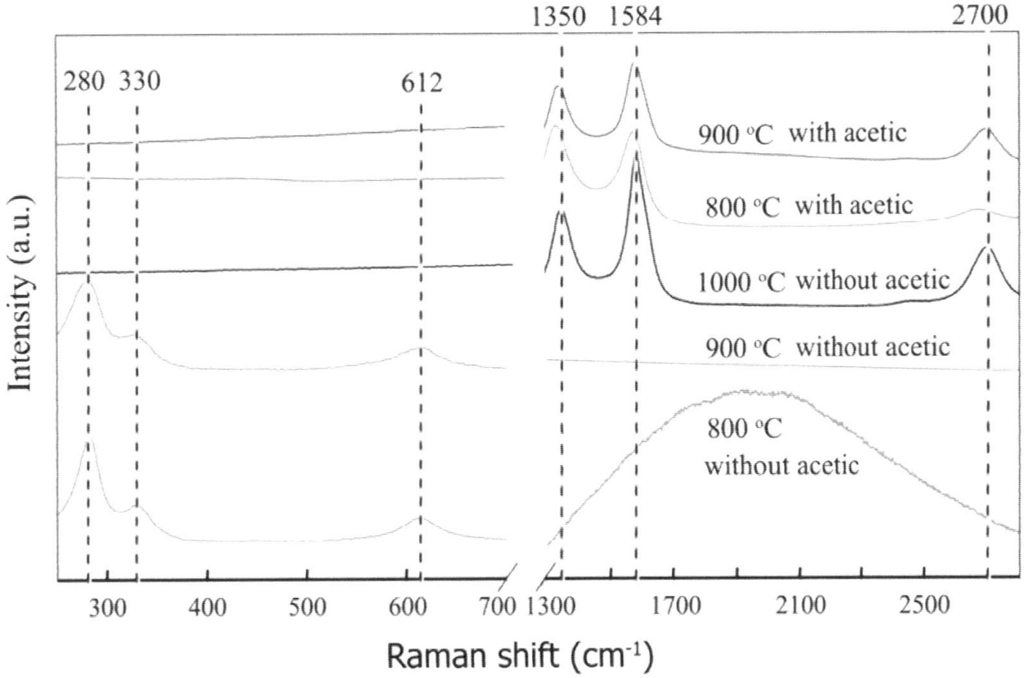

Figure 1. Raman spectra of as prepared copper with/without acetic pre-treatment at growth temperature from 800°C to 1000°C.

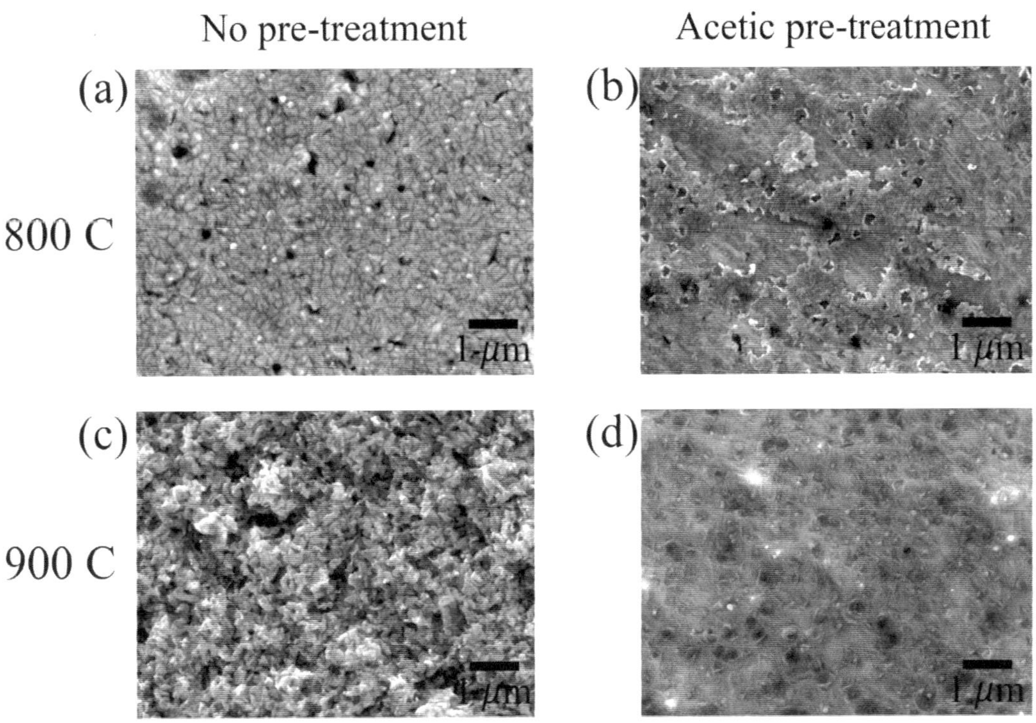

Figure 2. SEM images measured on the as prepared copper with/without acetic pre-treatment at annealing temperature of 800°C and 900°C.

Table 1. EDX analysis of as prepared copper with/without acetic pre-treatment at annealing temperature from 800°C to 1000°C.

Acetic pre-treatment	Growth temperature	Weight%			Atomic%		
		C	O	Cu	C	O	Cu
No	800	10.99	1.12	87.88	38.64	2.97	58.40
	900	4.00	19.35	76.65	12.12	44.00	43.88
	1000	63.52	-	36.48	90.21	-	9.79
Yes	800	18.97	0.66	80.37	54.73	1.43	43.83
	900	10.82	-	89.18	39.10	-	60.90

4. Conclusion

In this report, the effect of acetic acid pre-treatment on the growth of graphene on copper by CVD is studied. The acetic acid pre-treatment can eliminate copper oxide from the copper surface resulting in reduction of the growth temperature of graphene on copper.

5. References

[1] Hass J, Heer W A D and Conrad E H 2008 *J. Phys.: Condens. Matter* **20** 323202
[2] Lee C, Wei X, Kysar J W and Hone J 2008 *Sci.* **321** 385
[3] Nair R R, Blake P, Grigorenko A N, Novoselov K S, Booth T J, Stauber T, Peres N M R and Geim A K 2008 *Sci.* **320** 1308.
[4] Lee S M, Kim J H and Ahn J H 2015 *Mat. Today* **18** 336
[5] Lang F, Gluba M A, Albert S, Rappich J, Korte L, Rech B and Nickel N H 2015 *J. Phys. Chem. Lett.* **6** 2745
[6] Yazici M S, Azder M A, Salihoglu O and San F G B 2018 *International journal of hydrogen energy* **43** 18572
[7] Agostini M, Bruttu S and Hassoun J 2016 *ACS Appl. Mater. Interfaces* **8** 10850
[8] Singh R and Tripathi C C 2018 *Mater. Today: Proceedings* **5** 1125
[9] Hayes W I, Lubarsky G, Li M and Papakonstantinou P 2014 *Journal of Power Sources* **271** 312
[10] Ruammaitree A, Nakahara H, Akimoto K, Soda K and Saito Y 2013 *Applied Surface Science* **282** 297
[11] Hu H, Ruammaitree A, Nakahara H, Asaka K and Saito Y 2012 *Surf. Interface Anal.* **44**, 793
[12] Ruammaitree A, Nakahara H and Saito Y 2014 *Applied Surface Science* **307** 136
[13] Ruammaitree A, Nakahara H and Saito Y 2014 *Surf. Interface Anal.* **46**, 1156
[14] Ruammaitree A, Phokharatkul D, Nuntawong N and Wisitsoraat A 2018 *Surface Review and Letters* 1840003 DOI: 10.1142/S0218625X18400036
[15] Ruammaitree A, Phokharatkul D and Wisitsoraat A 2018 *Solid State Phenomena* **283** 173
[16] Mattevi C, Kim H and Chhowalla M 2011 *J. Mater. Chem.* **21** 3324
[17] Yoshihara N and Noda M 2017 *Chemical Physics Letters* **685** 40
[18] Rashad M, Rüsing M, Berth G, Lischka K and Pawlis A 2013 *Journal of Nanomaterials* **2013** 714853, http://dx.doi.org/10.1155/2013/71485
[19] Chavez K L and Hess D W 2001 *Journal of The Electrochemical Society*, **148** (11) G640

Acknowledgements

This work was supported by Energy conservation fund Thailand (contract no. FDA-CO-2560-3737-TH) and Research fund of Thammasat University (contract no. 6/2561).The authors thank Center of scientific equipment for advanced research, Thammasat University for SEM measurement.

Conductive paper of reduced graphene oxide and nanofibrillated cellulose

S Tanpichai[1,2,*], M Aachri[3,4], P Pattananuwat[3] and P Potiyaraj[3]

[1] Learning Institute, King Mongkut's University of Technology Thonburi, Bangkok, 10140, Thailand.
[2] Cellulose and Bio-based Nanomaterials Research Group, King Mongkut's University of Technology Thonburi, Bangkok, 10140, Thailand.
[3] Department of Materials Science, Faculty of Science, Chulalongkorn University, Bangkok, 10330, Thailand.
[4] University of Montpellier, Montpellier, 34090, France.

* Corresponding author: supachok.tan@kmutt.ac.th

Abstract. The conductive paper has been of great interest due to its flexibility and superior properties. In this work, the facile procedure with scalable possibility was used to prepare conductive paper with graphene oxide (GO) and reduced graphene oxide (rGO). GO synthesized by the modified Hummers' method was successfully transformed to rGO using ascorbic acid. These GO and rGO were used as a conductive source to improve conductive properties of the as-prepared paper. GO and rGO (0 – 25 wt%) were mixed with nanofibrillated cellulose (NFC) to form conductive paper. Results showed that the reduction of the electrical resistance was found from both GO/NFC and rGO/NFC paper sheets with increasing a content of GO or rGO. Furthermore, the paper with rGO exhibited more outstanding electrical conductive than the GO/NFC paper. Also, the water uptake of the conductive paper was affected by the addition of GO or rGO. The significant reduction of water uptake was found to be more pronounced with the introduction of rGO in comparison with the presence of GO. The conductive paper prepared in this study would be useful for applications such as flexible electrode.

1. Introduction

Flexible electronic devices have been of great interest due to a wide range of applications such as portable electronics, energy storage, photovoltaic cells and smart clothing [1, 2]. Due to environmental awareness, bio-based materials with advantages of light weight, biodegradability and superior mechanical properties have attracted much attention [1, 2]. Nanofibrillated cellulose (NFC), a web structure of nanofibers and microfibers entangled, has been named by Herrick *et al.* [3] and Turbak *et al.* [4]. Later, cellulose fibers with diameters in the range of 10 – 100 nm were finally obtained using several approaches, such as grinding [5], high speed blending [6] or the combination of high speed homogenizaton and sonication [7-9]. NFC could be used as an alternative raw material to traditional conductive rigid and fragile films like ITO [2].

To improve conductive properties of cellulose, a conductive material is required. In comparison with other conductive materials such as carbon nanotube, conductive polymer and metallic nanowire, two-dimensional-structured graphene has been proved to have greater potential due to its high surface area, superior mechanical properties and biocompatibility [1, 10]. Recently, the manufacture of graphene in a large production could be feasible via graphene oxides (GO) with the introduction of the reducing agent to remove functional groups such as hydroxyl, carboxyl and carbonyl [1, 11].

In this research, flexible conductive NFC paper with GO and rGO was prepared. Water uptake and electrical resistance of the prepared conductive paper were investigated.

2. Materials and methods

2.1. Materials and chemicals

Nanofibrillated cellulose (NFC, CELISH KY100S) was supplied by Daicel Finechem Co., Ltd. Graphite powder, sodium nitrate ($NaNO_3$), potassium permanganate ($KMnO_4$), hydrogen peroxide (H_2O_2), sulfuric acid (H_2SO_4), hydrochloric acid (HCl) and ascorbic acid were purchased from Sigma-Aldrich Co., LLC.

2.2. Preparation of graphene oxide

Graphene oxide (GO) was synthesized by the modified Hummers' method as described elsewhere [12]. Briefly, graphite flakes (10 g) and $NaNO_3$ (5 g) were slowly added into a concentrated H_2SO_4 solution at 0 °C. Then, 30 g of $KMnO_4$ was used to oxidize the suspension at the temperature below 20 °C. The suspension was further stirred for 120 min at 0 °C, and continue kept stirring at 35 °C for 30 min. After that, 360 ml of distilled water was slowly added to the suspension. At this state, the temperature of the suspension was rapidly increased to ~100 °C due to an exothermic reaction. The suspension was further maintained at 80 °C for 15 min, and was subsequently diluted with 1,140 ml of distilled water, and 10 ml of H_2O_2. Finally, the GO suspension was filtered, and washed with HCl, and distilled water. The obtained GO cake was dried at 70 °C for 48 h. A GO dispersion (5 mg ml^{-1}) was prepared with the aid of ultrasonic bath (50 Hz) for 3 h, and followed by centrifugation at 3,000 rpm for 15 min to remove the un-exfoliated GO. The reduction of GO to rGO was carried out before the conductive paper preparation. 1 g of GO was soaked in the ascorbic acid solution (4 wt%) under stirring for 24 h.

2.3. Preparation of conductive paper of graphene oxide and nanofibrillated cellulose

0.5 wt% NFC suspension was prepared using a high speed homogenizer (Ultra Turrax IKA-T25) for 30 min. Then, a GO or rGO suspension with a concentration of 0 - 25 wt% was slowly added into the NFC suspension. This suspension was stirred for 60 min. The NFC paper with GO or rGO was formed by a filtration method to obtain a wet cake. Subsequently, the wet cake was oven-dried at 40 °C for 48 h under a mass of 4 kg, and was further dried using a compression molding with the pressure of 100 bar at 110 °C for 30 min. The appearance of the NFC/GO paper is illustrated in Figure 1.

2.4. Characterizations

2.4.1. Fourier transform infrared spectrometry (FTIR). Chemical functional groups of GO and rGO sheets were investigated using a Fourier transform infrared spectrometer (Thermo Nicolet 6700) equipped with an attenuated total reflectance (ATR) mode in a range of 4000 – 500 cm^{-1}. Sample spectra were recorded with a resolution of 4 cm^{-1} and a repetitious scan of 64 times. The background was collected before each test.

2.4.2. Water uptake. Prior to the measurement, all samples were dried in an oven at 60 °C for 48 h to obtain the constant weight. The samples were subsequently soaked in distilled water for specific times (2, 4, 6 and 8 h) at room temperature. After that, the sample was taken out, wiped out water with filter paper, and immediately weighted. The water uptake of the samples (W) was calculated using the equation.

$$W(\%) = ((W_t - W_i)/W_i) \times 100 \quad (1)$$

where W_t and W_i are the weight of the sample at various soaking times and the initial dried weight of the sample, respectively.

2.4.3. Electrical resistance. The electrical resistance of the samples were measured using an ohmmeter. The standard positive and negative probes were attached on samples, and a value was recorded.

3. Results and discussion

Figure 2 presents FTIR spectra obtained from GO and rGO materials. The FTIR spectrum of the GO was more characteristic than that of rGO. The outstanding peak located at 3340 cm^{-1}, attributed to OH stretching vibration could be easily observed from GO [13, 14]. The peaks at 1720 cm^{-1} could be assigned to C=O stretching of carboxylic and/or carbonyl moiety functional groups. The presence of the 1049 cm^{-1} peaks indicated the stretching vibration of the C-O-C groups. It would be noted that some of these peaks could not be detected from the rGO material, and the intensity of the remaining peaks was significantly lower in comparison with the GO spectrum. This was an indication of the less hydroxyl groups remained on the rGO sheets after the reduction with the ascorbic acid.

Figure 1. Appearance of the GO/NFC paper

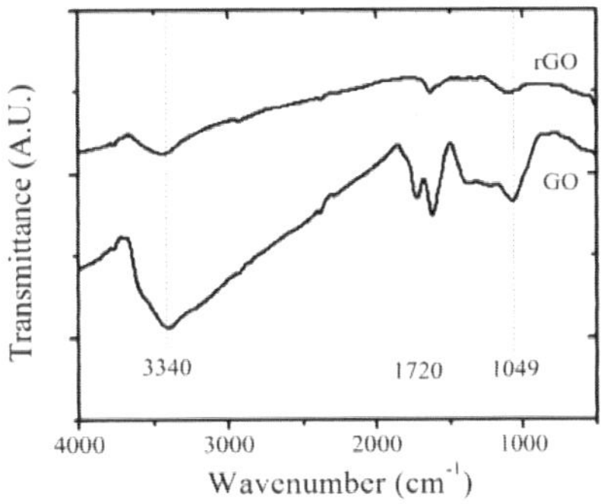

Figure 2. FTIR spectra obtained from rGO and GO.

Figure 3 presents the water uptake of the conductive paper with GO and rGO. Water uptake of the NFC paper without the addition of GO and rGO presented a water uptake of 45.0 %. No significant change of the water uptake for the GO/NFC paper was found when the GO content was higher to even 25 wt%. This might be due to the oxygen functional groups on GO sheets interacted with water. On the other hand, with introduction of the rGO in the NFC paper, the decreased water uptake was monitored. The water uptake of the paper with 5 wt% rGO decreased to 32.2 % after soaking in distilled water for 8 h. When the percentage of rGO in the paper was higher, the water uptake decreased slightly. At 25 wt% of rGO in the paper, its water uptake was 24.5 %. This reduction was due to the hydrophobic rGO (less functional groups interacted with water) and a lower amount of cellulose [14]. With a higher content of rGO, the more hydrophobic paper was formed. This was well agreement with the study of Mianehrow *et al.* [1] reporting that the hydrophobicity of the composites increased with increasing a quantity of rGO.

The electrical resistance of the paper as a function of the GO or rGO content is presented in Figure 4. In this study, the highest electrical resistance could be obtained from the paper without the presence of GO and rGO. With 4 wt% of GO, the electrical resistance was 771 kΩ, and this was significantly reduced to 19.5 kΩ when the GO content increased to 10 wt%. After 10 wt% of GO, the electrical resistance of the paper decreased slightly. The electrical resistance of 2.6 kΩ was obtained from the paper with 25 wt% of GO. On the other hand, the reduction of GO to rGO could increase the conductivity [1]. A value of 0.78 kΩ could be measured from the paper with 5 wt% rGO. With increasing a content of rGO, the electrical resistance decreased considerably. At 25 wt% of rGO, the paper showed the electrical resistance of 0.038 kΩ. This high conductivity indicated that the less-contaminated and damage rGO sheets were well-dispersed in the NFC paper, allowing electrical charges to be efficiently transferred. The similar reduction of the electrical resistance has been reported for the recycled pulp mixed with rGO [1].

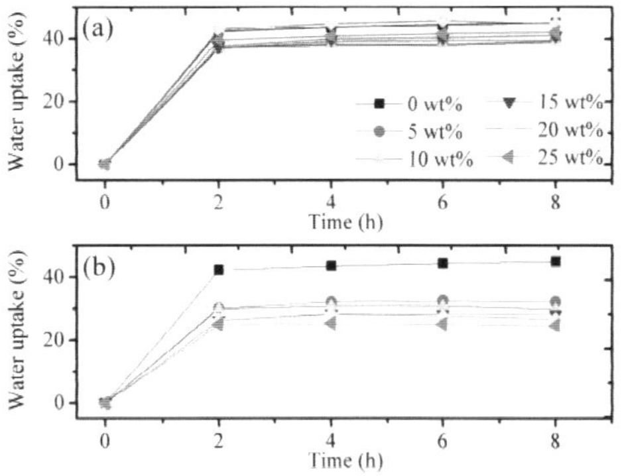

Figure 3. Water uptake of the NFC paper with (a) GO and (b) rGO with immersion times.

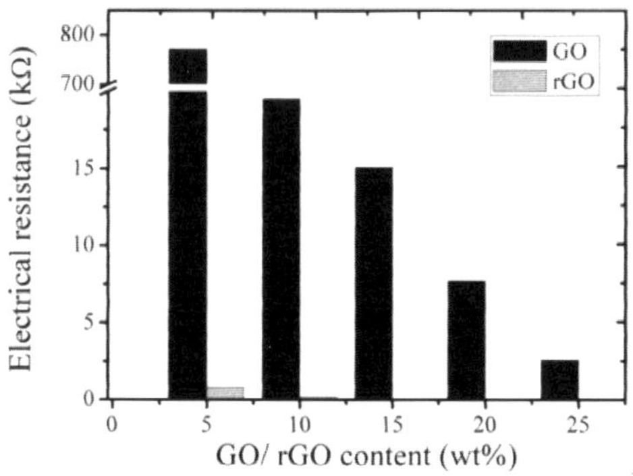

Figure 4. Electrical resistance of the NFC paper with GO and rGO.

4. Conclusions

Flexible conductive paper sheets of NFC with two different types of graphene sources (GO and rGO) were successfully prepared. With the introduction of ascorbic acid, GO was reduced to rGO. Due to the 2D-honeycomb structure of graphene sheets with less oxygen functional groups on the surface of rGO, the conductive paper with rGO showed higher electrical conductivity and lower water absorption compared to the GO/NFC paper. With increasing a GO quantity, no significant change of water uptake could be observed from the paper, and the electrical resistivity of the paper decreased considerably with the addition of GO. The NFC paper with 5 wt% GO showed the electrical resistivity of 771 kΩ while the electrical resistivity value of 19.5 kΩ was measured from the NFC paper with 10 wt% of GO. On the other hand, water uptake of the NFC paper with rGO decreased greatly when the rGO content was higher. With 5 wt% of rGO, the electrical resistivity of the NFC paper was plummeted to 0.78 kΩ, and the electrical resistivity decreased to 0.038 kΩ when the rGO content was 25 wt%. We believed the prepared conductive paper of rGO and NFC could be potentially used for low cost and environmental friendly flexible electronic devices.

5. References

[1] Mianehrow H, Sabury S, Bazargan A M, Sharif F and Mazinani S 2017 *J. Mate. Sci. Mater. Electron.* **28** 4990
[2] Gao K, Shao Z, Wu X, Wang X, Li J, Zhang Y, Wang W and Wang F 2013 *Carbohydr. Polym.* **97** 243
[3] Herrick F W, Casebier R L, Hamilton J K and Sandberg K R 1983 *J. Appl. Polym Sci: Appl Polym Symp* **37** 797
[4] Turbak A F, Snyder F W and Sandberg K R 1983 *J. Appl. Polym Sci: Appl Polym Symp* **37** 815
[5] Iwamoto S, Nakagaito A N and Yano H 2007 *Appl. Phys. A-Mater. Sci. Process.* **89** 461
[6] Uetani K and Yano H 2011 *Biomacromolecules* **12** 348
[7] Tanpichai S, Sampson W W and Eichhorn S J 2012 *Compos. Pt. A-Appl. Sci. Manuf.* **43** 1145
[8] Tanpichai S, Sampson W W and Eichhorn S J 2013 *Adv. Mater. Res.* **747** 359
[9] Tanpichai S, Sampson W W and Eichhorn J S 2014 *Compos. Pt. A-Appl. Sci. Manuf.* **65** 186
[10] Ahmed J, Mulla M, Arfat Y A and Thai T L A 2017 *Food. Hydrocolloid.* **71** 141
[11] Sadasivuni K K, Kafy A, Kim H-C, Ko H-U, Mun S and Kim J 2015 *Synth. Met.* **206**
[12] Hummers W S and Offeman R E 1958 *J. Am.Chem. Soc.* **80** 1339
[13] Rattana, Chaiyakun S, Witit-anun N, Nuntawong N, Chindaudom P, Oaew S, Kedkeaw C and Limsuwan P 2012 *Procedia. Eng.* **32**
[14] Etmimi H M, Mallon P E and Sanderson R D 2013 *Eur. Polym. J.* **49** 3460

IOP Conf. Series: Materials Science and Engineering **526** (2019) 012010 doi:10.1088/1757-899X/526/1/012010

Material Ratio Analysis of Charcoal Briquettes from Dendrocalamus Asper Backer Bamboo

R Lukmuang[1], C Dasaard[1], P Chantawong[1], D Ngamrungroj[1,*]
[1]Energy Engineering Technology, Department of Power Engineering, King Mongkut's University of Technology North Bangkok, Bangkok, Thailand

*Corresponding author: nuuoeiz.a1@gmail.com

Abstract. The objective of this study proposes the best ratio of charcoal briquettes from 'Dendrocalamus Asper Backer' bamboo to make a comparison between heating value and environmental factors. A sample was selected from proportion of charcoal powder and cassava powder with 100:0, 95:5, 90:10, 85:15, 80:20, 60:40, 50:50. The proportion was used for data analysis using temperature machine, thermal gravimetric analyzer (TGA) and gross calorific value as ASTM 5865-04 standard. The results of present study were summarized as follows: 1) the best ratio is 90:10 with heating value 7419 cal/g, 2) higher charcoal powder higher heating value, 3) higher cassava powder higher smoke and soot.

1. Introduction

Thailand is an agricultural country with plenty of biomass [1]. Variety of material resources is used to produce energy such as sugar cane, straw, napier grass, or even agricultural waste. However, these materials are not enough for the energy requirements. Therefore, it has to find new potential resources of renewable energy with sufficient quantities.

Concerning to the demand of biomass fuels is increasing, the purpose of this study is to find suitable biomass resources for easy growing terrain and high potential for biomass production. Bamboo, from 'Dendrocalamus Asper Backer', can be applied for energy production.

The research study found that bamboo is the fastest growing plant in the world. This can grow at an average of 10 cm/day [2]. This is the main plant for production of environmentally friendly products, easily grown once cultivated for 70-80 years without replanting, high density, drought resistance and high heating value when compare with other local biomass materials [3-5]. Therefore, the researcher will find the optimal ratio for using as charcoal briquettes.

2. Research Methodology

2.1 Material

The main material is bamboo (Dendrocalamus Asper Backer) planting at "Jiraphan Bamboo Garden", Yasothorn, Thailand. Average age is 0-5 years for main material and combust with 200 liters tank for increase high heating of material. Pyroligneous acid or wood vinegar is a by-product of charcoal burning that is the crude condensate of smoke generated during the process of making wood charcoal [5].

2.2 Charcoal to Charcoal Briquettes

1) Mash machine with 3 HP grinder hammer mill is used to mash charcoal into charcoal powder in order to give high heating and decrease smoke when it was burned.
2) Mixing charcoal powder and cassava powder with 5 HP mixer machine in ratios of 100:0, 95:5, 90:10, 85:15, 80:20, 70:30, 60:40, and 50:50.
3) Compressing charcoal powder and cassava powder with 5 HP conical screw press machine using standard diameter size 10 cm long and 1 cm hole.
4) Drying charcoal briquettes after compress to reduce moisture, increase heating value and make it hardly broken.

2.3 Experiment

Cutting 3 pieces of briquettes into 5 cm long and 1 cm diameter, and then heating samples using stove in 2 min to control heating value and time. After that, heating samples in 1000 ml of water for 3 h to find acceleration of heat and time of each ratio without addition of oxygen in combustion.

2.4 Performance Measurement

Analysis by temperature measuring sensors connect to computer as shown in figure 1.

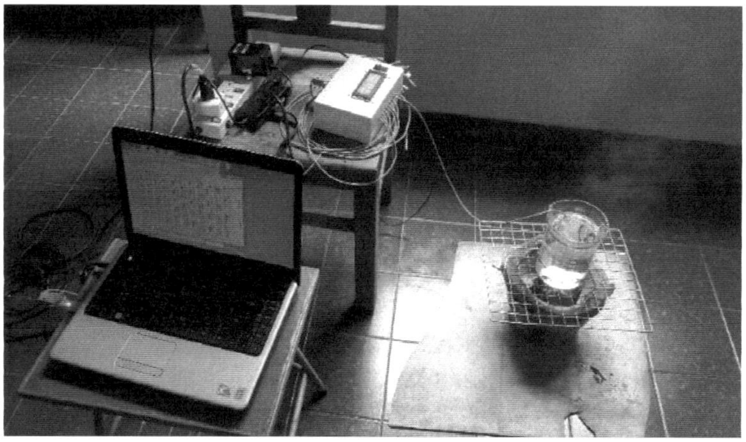

Figure 1. Primary experiment and measurement

3. Results

We found that the best ratio of charcoal powder and cassava powder is 90:10 with high heating value, less smoke and smut. However, the content of cassava powder over 40 percent affects to stuck in output hole of machine. High heating acceleration belongs to charcoal powder proportion as shown in figure 2, combustion lifetime depends on cassava powder proportion at the expense of heating value.

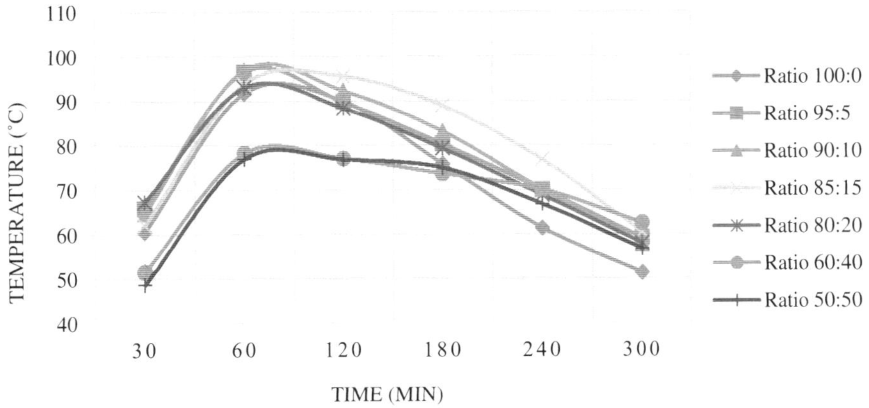

Figure 2. Heating comparison of samples

Fuel characteristics of each ratio in experiment present percentage of moisture, ash, volatile matter, fixed carbon and heating value (Cal/g) as shown in table 1. The best ratio is 90:10 compared to other local material briquettes in Thailand as shown in table 2 [6].

Table 1. Approximate analysis and heating value in experiment.

Ratio proportion	Moisture (%)	Ash (%)	Volatile Matter (%)	Fixed Carbon (%)	Heating Value (Cal/g)
Ratio 100:0	3.512	4.504	21.160	70.824	7,518
Ratio 95:5	2.830	3.215	24.371	69.584	7,408
Ratio 90:10	2.015	2.591	26.479	68.915	7,419
Ratio 85:15	2.005	2.614	27.488	67.893	7,389
Ratio 80:20	1.982	2.324	30.382	65.312	7,201
Ratio 60:40	1.975	2.412	35.402	60.211	6,918
Ratio 50:50	2.013	2.531	37.263	58.193	6,549

Table 2. Comparison between experiment and other material briquettes.

Biomass Type	Moisture (%)	Ash (%)	Volatile Matter (%)	Fixed Carbon (%)	Heating Value (Cal/g)
Dendrocalamus Asper Backer bamboo	5.8	2.7	71.7	19.8	4,203
Charcoal briquettes (Dendrocalamus Asper Backer bamboo) (Ratio 90:10)	2.015	2.591	26.479	68.915	7,419
Palm shell	12.12	3.66	68.31	16.3	4,413
Charcoal briquettes (Coconut shell)	7.13	3.74	13.47	82.79	7,276
Rice husk	12.5	12.73	56.98	18.88	3,499
Rice straw	10.12	10.42	60.87	18.8	3,173
Bagasses	50.76	1.75	41.99	5.86	2,308

Biomass Type	Moisture (%)	Ash (%)	Volatile Matter (%)	Fixed Carbon (%)	Heating Value (Cal/g)
Rubberwood	-	2.1	74.9	23	4,560
Charcoal briquettes (Rubberwood)	-	3.4	17.5	79.1	7,650
Charcoal briquettes (Rambutan wood)	3.415	3.932	21.815	70.838	7,059
Coal	-	7.7	42.8	49.5	5,860
Coke (fuel)	-	8.2	1.2	90.6	7,150

4. Conclusions

Group of charcoal briquettes ratio of 100:0, 95:5, 90:10, 85:15, 80:20 has better heating acceleration than those of 60:40 and 50:50 ratio. Actually, the proportion of maximum heating ratio should be the ratio with the highest number of charcoal powder. However, the 100:0 ratio experiment did not exhibit maximum heating during charcoal combustion. The lowest heating ratio in experiment was 50:50 as expected. The ratio of charcoal powder affects heating duration when cassava powder content is higher because cassava powder is burned slower than charcoal powder. Higher cassava powder content gives higher smoke and soot. During burning, the 100:0 ratio between charcoal powder and cassava powder has a bit smoke and soot but the 50:50 ratio has a lot of smoke and soot at the bottom of beaker. Soot color is dark brown and sticky.

5. References

[1] Yotakong S. 2010 Thai Society and Management, Promotion and Development of Agriculture
[2] U.S. Departmane of Energy 2000 Bamboo: an overlooked biomass resource?
[3] Truong A.H., Le T.M.A 2014 Overview of bamboo biomass for energy production
[4] Scurlock, J.M.., Dayton. D., Hames. B. 2000 Bamboo: an overlooked biomass resource?
[5] Sritong, C., Kunavongkrit, A., Piumsombun, C. 2012 Bamboo: An innovative alternative raw material for biomass power plant. Int. J. Ò Innov. Manag. Technol. 3.
[6] Thailand Institute of Scientific and Technological Research, Sustainable Development
[7] GE Technology inc., Magic Power of Bamboo Charcoal/Activated Carbon
[8] Shenxue J. 2014 Training Manual of Bamboo Charcoal for Producers and Consumers
[9] Singh G., Richa, M.L. Sharma 2017 Bamboo – A Miracle Plant
[10] Piriyayotha T. 2018 The Design of Briquetting Machine for Biomass Residues Fuel
[11] Sajjakul K., Hengniran P., Neimsuwan T., Leamsak N. 2016 A production of high quality bamboo charcoals as activated charcoal based materials using 200 liters oil drum modified charcoal kiln
[12] Phutteesakul R. 2010 The Production of Charcoal Briquette by Coconut Shell and Cassava Rhizome
[13] Torsakul S., Thongsri K., Supharattana C. 2012 Development of Charcoal Briquette from Scrapped Coconut for Alternative Energy
[14] Sumanatrakul P., Kongsune P., Chotitham L., Sukto U. 2015 Utilization of Dendrocalamus Asper Backer Bamboo Charcoal and Pyroligneous Acid
[15] Malanit P. 2009 The Suitability of Dendrocalamus asper Backer for Oriented Strand Lumber

IOP Conf. Series: Materials Science and Engineering **526** (2019) 012011 doi:10.1088/1757-899X/526/1/012011

Adhesion and heat build-up of rubber for energy-saved tyre

S Suethao[1] and W Smitthipong[1,2,*]

[1] Specialized center of Rubber and Polymer Materials in agriculture and industry (RPM), Department of Materials Science, Faculty of Science, Kasetsart University, Bangkok, Thailand.
[2] Office of natural rubber research program, The Thailand Research Fund, Bangkok, Thailand.

*Corresponding author: fsciwssm@ku.ac.th

Abstract. The objective of this study was to investigate the effect of crosslinking systems (conventional and efficient systems) on the contact mechanics and heat build-up of natural rubber (NR) tyre mimicking samples. The JKR (Johnson, Kendall and Roberts) experiment has been established to measure the contact mechanics in NR samples at different vulcanization systems. Work of adhesion obtained from JKR model is in good agreement with that calculated from contact angle measurement. From the contact mechanics point of view, the adsorption of the rubber to the glass slide surface plays a major role in the strain energy release rate. Higher rubber free chain samples (unvulcanized rubber) represent higher adhesion hysteresis and higher heat build-up compared to vulcanized rubbers whichever was the vulcanization system. Finally, rubber sample obtained using efficient vulcanization has the lowest adhesion hysteresis and heat build-up. These results suggest that the rubber with efficient vulcanization is preferable for the application of energy-saving tyres.

1. Introduction

Natural rubber (NR) was widely used as a viscoelastic material in many types of product [1-2]. Based on the global demand of energy-saved products, understanding of contact mechanics and vulcanization system of NR is very important. When a hemispherical rubber substrate contacts with another flat object, the adhesion forces acting across the interface tend to deform the solids and thus increase their contact area. At its simplest level, the experimental methodology involves bringing a deformable hemispherical object into contact with a flat substrate under controlled loads. Mechanical calibration of the load-deformation data using the equation yields the adhesion energy [3]. The Johnson-Kendall-Roberts (JKR) contact mechanics test is a simple test based on contact mechanics that has been more fully described elsewhere [4-5]. The JKR theory modifies the classical Hertz [6] result for the deformation of a sphere in contact with a flat surface to include the effect of interfacial energy or work of adhesion; the contact radius, a, is related to the applied load, P, by

$$a^3 = \frac{R}{K}\left[P + 3\pi WR + \sqrt{6\pi WRP + (3\pi WR)^2}\right] \quad (1)$$

IOP Conf. Series: Materials Science and Engineering **526** (2019) 012011 doi:10.1088/1757-899X/526/1/012011

where R is the radius of the undeformed sphere, K the elastic modulus and W the work of adhesion of the hemispherical polymer and the substrate. A typical JKR experiment is performed in three steps. The first stage is known as the loading process, which involves compressing the hemispherical elastomer against a flat surface under controlled loads. By determining the variation of the contact radius with applied load, it is possible to fit such data to equation (1) and to extract two independent properties K and W [4]. In the second stage, the elastomer remains in contact with the substrate for a fixed period of time. During this time interval, the evolution of physical and chemical interactions may lead to an enhancement in the adhesion energy. The final stage, or the unloading process, entails rapidly decompressing the hemispherical elastomer from the flat surface. One can then determine the strain energy release rate, G, necessary to maintain a certain radius of contact, a, between a flat surface and an elastic hemisphere of radius R under a contact force, P, by

$$G=\frac{\left(P-\frac{Ka^3}{R}\right)^2}{6\pi Ka^3}=\frac{\left(P-P_H\right)^2}{6\pi RP_H} \tag{2}$$

where $P_H = Ka^3/R$ is the Hertz contact pressure, the pressure that would be required to produce the measured a if there were no adhesive forces ($G = 0$). If equilibrium loading and unloading conditions are achieved, and if there is an absence of interface reconstruction, G should equal W, the equilibrium work of adhesion. In the case of two identical surfaces in contact, W is defined as twice the surface energy of that material [7]. The contact mechanics relates to the hysteresis of NR products, thus the lost energy of materials can be investigated [3,7]. The objective of this study is to investigate the effect of vulcanization system on the contact mechanics and vulcanization system of NR samples.

2. Experimental

Thai rubber block (STR 20) is used as a raw material for preparation of NR samples. The composition of rubber samples is as stated in table 1. Samples were prepared with internal mixer (type kneader) and two-roll mill.

Table 1. Formulas of rubber compound.

Formula (phr)	Control	Conventional Vulcanization (CV)	Efficient Vulcanization (EV)
NR (STR20)	100	100	100
ZnO	-	3	3
Stearic acid	-	2	2
TBBS	-	1	3
Sulphur	-	3	1
Total	100	109	109

Three types of hemispherical rubber compounds (control, CV and EV) were prepared in a hemispherical mould of 1 cm diameter using the procedure described elsewhere [3] for the study of contact mechanics using JKR experiment. The surface energies of the different rubbers and glass slide were evaluated from contact angle measurements using different liquid drops of known properties [2]. Tensile test is one of the most common tests to study mechanical properties of rubber. For the tensile test, rubber samples were cut in a dumbbell shape. The rate of testing crosshead speed was 500 mm/min at room temperature. Each material of the rubbers was tested at least three times. Crosslinking density of rubber networks is measured by its swelling, the number of effective network chains per unit volume of rubber can be calculated according to the Flory-Rehner equation [8]. Heat build-up was measured by BF Goodrich Flexometer (Model II) following ASTM D623 method of measurement.

3. Results and discussion

Table 2 shows that NR compound of CV trends to have lower crosslinking density than that of EV in good agreement with the tensile strength from tensile test. On the other hand, NR compound of CV is better in elongation than that of EV. NR with conventional vulcanization (CV) represents long sulphur crosslinking bridge compare to NR with efficient vulcanization (EV), longer sulphur bridge better elongation. NR samples with both CV and EV clearly represent better mechanical properties (tensile strength and elongation at break) than unvulcanized NR (control).

Table 2. Properties of NR compounds.

Properties	**Control**	**CV**	**EV**
Crosslinking density (mole/m^3)	-	105 ± 3	110 ± 3
Tensile strength (MPa)	0.13 ± 0.01	23.95 ± 1.20	28.45 ± 1.42
Elongation at break (%)	248 ± 25	1826 ± 183	1383 ± 138

The surface energy of natural rubbers and glass slide can be determined using contact angle measurements. There are no significant differences of surface energy of all NR samples (below 50 mJ/m^2 means non-polar materials). For the glass slide, it represents as a polar material (above 50 mJ/m^2). Concerning to the work of adhesion [2] between rubber samples and glass slide, there are no significant differences of work of adhesion from contact angle measurement: control/glass, CV/glass and EV/glass (table 3). Based on the contact mechanics test, a load and unload experiment was carried out to examine the contact mechanics of the hemispherical rubber [8]. The applied load (P) was increased stepwise to the final load and then released stepwise to the initial load. At each load, the radius of the contact zone (a) between the hemispherical rubber and glass slide was measured. Concerning to the relationship between a^3 and P for the loading experiment, typical results obtained during the loading experiment were fitted to equation (1) to extract work of adhesion (W). We found that work of adhesions obtained from JKR model are in good agreement with those calculated from contact angle measurement (table 3).

Table 3. Work of adhesion (W) of all NR samples.

Work of adhesion (mJ/m^2) ± 10%	**Control/glass**	**CV/glass**	**EV/glass**
Calculation from contact angle measurement	97.32	98.08	99.28
Extraction from JKR model	90.13	91.95	92.45

A typical JKR experiment is used in this study, it consists of bringing into contact two samples in the shape of a semi-spherical NR and a glass slide under the influence of an external load that increases until maximum load is reached and subsequently retracting them till the contact is broken. The adhesion hysteresis of PDMS networks has also been reported in the literature [4]. It is therefore of interest to investigate such adhesion hysteresis in our system because it should be related in some instances to penetration of chains into the networks and in others to interface reconstruction giving at the interface. To explore the adhesion hysteresis provides by the rubber samples, we have prepared three types of rubber samples: unvulcanized NR, NR with CV and NR with EV. Figure 1 shows the results of the contact mechanics between different rubber samples with glass slide, analyzed with equation 2 to find G as a function of a. We found that all the samples displayed adhesion hysteresis (different strain energy released rate) between the loading and unloading experiments. However, the amount of that hysteresis increased significantly on going from: NR with EV < NR with CV < unvulcanized NR.

The strain energy release rate (G) measured during the loading experiment (closed symbols) remains independent of a, in good agreement with the expected value of twice the surface energy of control NR (32.0 mJ/m^2). However, the unloading G actually increases as a decrease from the maximum contact radius (corresponding to the maximum load). As the length and flexibility of sulphur crosslinking bridge is increased (CV > EV) the adhesion hysteresis and heat build-up also increase (CV > EV). The vulcanization system of NR samples plays a major role in the adhesion hysteresis and heat build-up. Moreover, unvulcanized NR represents higher adhesion hysteresis and heat build-up than vulcanized NR. This result implies that the mobility of macromolecular chains is an important factor in determining the hysteresis of rubber chain at the interface of substrate and the heat build-up of rubber. The rubber sample with EV should be applied for the energy-saved tyre.

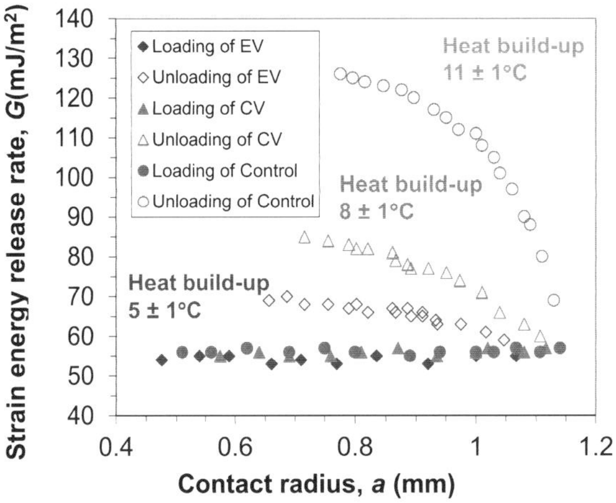

Figure 1. Loading (closed symbols) and unloading (opened symbols) G as a function of contact radius, a, contact with glass slide for three types of rubber samples: unvulcanized NR, NR with CV and NR with EV.

4. Conclusions

The JKR experiment has been well-developed to study the contact mechanics of NR samples with respect of vulcanization system. We found that the adsorption of rubber chains on the surface of glass slide plays a major role in the hysteresis phenomenon. This result indicates a strong dependence of hysteresis and heat build-up on the presence of rubber free chains.

5. References

[1] Smitthipong W, Nardin M, Schultz J and Suchiva K 2007 *Inter. J. Adhesion Adhesives* **27** 352.
[2] Smitthipong W, Suethao S, Shah D and Vollrath F 2016 *Polym. Testing* **55** 17.
[3] Johnson K L, Kendall K and Robert A D 1971 *Proc. R. Soc. London: Part A* **324** 301.
[4] Falsafi A, Bates F S and Tirrell M 2001 *Macromolecules* **34** 1323.
[5] Perutz S, Kramer E J, Baney J and Hui C Y 1997 *Macromolecules* **30** 7964.
[6] Hertz H J 1896 In Miscellaneous Papers, *Macmillan & Co.*, London.
[7] Zeng H, Tian Y, Boxin Z, Tirrell M and Israelachvili J 2007 *Macromolecules* **40** 8409.
[8] Flory P J 1950 *J. Chem. Phys.* **18** 108.

Acknowledgement
This work was funded by the Thailand Research Fund.

Study of the effect of electron beam on thermal and mechanical properties of poly(lactic acid)/poly(butylene succinate) blends

S Tiptipakorn[1,*], K Hemvichain[2], and A Pirombua[1]

[1] Department of Chemistry, Faculty of Liberal Arts and Science,
Kasetsart University, Kamphaeng Saen Campus, Nakhon Pathom 73140, Thailand.
[2] Thailand Institute of Nuclear Technology (Public Organization),
Ministry of Science and Technology, Ongkharak, Nakhon Nayok 26120, Thailand.

* Corresponding author: faassntk@ku.ac.th

Abstract. In this research, the effect of radiation dose on thermal and mechanical properties of polymer blends between poly(lactic acid) (PLA) and poly(butylene succinate) (PBS) were determined at the PLA/PBS weight ratios of 100/0, 80/20, and 0/100. Dicumyl peroxide (DCP) was used as additive at 0.2 phr. The radiation doses were varied at 0-150 kGy. There were some changes of chemical structure upon irradiation as observed via Fourier Transform Infrared Spectrometer (FT-IR), leading to the decrease of melting temperature (T_m) with the increase of radiation doses. Moreover, the flexural strength determined via Universal Testing Machine (UTM) were decreased with increasing radiation doses. The lowest biodegradation was found at the radiation dose of 50 kGy.

1. Introduction

The use of plastic from petroleum in the recent years has led to serious environment pollution due to their non-biodegradability. Biodegradable plastic has attracted most interest because of more environmentally awareness. Especially, biodegradable polyester contains numerous ester linkage which can be easily hydrolysed by water. Poly(lactic acid) (PLA), poly(butylene succinate) (PBS), polycaprolactone (PCL) and polyhydroxyalkanoate (PHA) are such well known as excellent biodegradability aliphatic polyesters.

PLA has been widely used to substitute conventional plastic due to various outstanding characteristics for example great tensile modulus and flexural modulus over high density polyethylene (HDPE), polypropylene (PP) and polystyrene (PS). However, low elongation at break and impact strength of PLAs are their main drawbacks [1]. Moreover, PLAs have low thermal and dimensional stabilities under high temperature. To improve those mentioned properties, the crosslinking technique has been applied. PBS is also one kind of aliphatic biodegradable polyester to as PLA. PBS has gained much interesting due to its good impact strength, high flexibility and excellent thermal properties (degradation temperature of 399°C). The applications of PBS include food container, plastic bottle and food-graded film. PBS is widely used to improve properties by blending or blend to other polymers.

The blend of PLA/PBS has been developed using dicumyl peroxide (DCP) as radical initiator using reactive blending [2]. By this procedure, it is believed that the partial miscible blend of the PLA/PBS occurred as evidenced of some degree of crosslinking and branching between PLA and PBS molecules. Moreover, it was reported that the DCP content of 0.2 and 0.3 phr provided the greatest properties of PLA/PBS.

The objective of this research is aimed to develop mechanical and thermal properties of PLA by blending with high flexibility and thermal stability PBS using DCP as compatibilizer. The effects of various PLA/PBS blending weight ratios and radiation doses on thermal and mechanical properties of PLA/PBS blend were investigated.

2. Experiment

2.1 Materials and polymer preparation

Raw materials used in this research were PLA and PBS. PLA (Ingeo Biopolymer 3052D) having molecular weight of 180,000 dalton was supplied by NatureWorks. PBS (BioPBSTM FZ91PM, density of 1.26 g cm^{-3}) and dicumyl peroxide (98% AR grade) were purchased from Sigma Aldrich Co.

2.2 Preparation of the polymer blend

Both PLA and PBS pellets were initially dried in an air-circulated oven at a temperature of 50°C for 2 hours to remove moisture before processing and kept in a desiccator at a room temperature. The PLA and PBS having weight ratios of 100/0, 80/20 and 0/100 were then mixed with 0.2 phr of DCP via an internal mixer at a temperature of 170°C for 10 minutes. The mixed polymer was hot pressed into a test specimen using a hydraulic compression molder (model PR1D-W300L300 HD). An electron beam radiation was carried out at various radiation doses of 0, 50, 100 and 150 kGy.

2.3 Characterization of the polymer blend

Chemical structure and network formation of a sample was studied by Fourier transform infrared spectroscopy (FTIR). Fourier transform infrared spectra of all samples were acquired by a FT-IR spectrometer from Bruker Tensor 27. All spectra were taken as a function of time with 512 scans at a resolution of 4 cm-1 and a spectral range of 2000-800 cm^{-1}.

A differential scanning calorimeter (DSC) model DSC822e from Mettler Toledo, was used to study melting temperature (T_m) of the PLA/PBS blend. All samples were put in 40 μl aluminum pans with lids. The sample with a mass in a range of 5-10 mg was sealed in an aluminum pan. The DSC thermogram was obtained using a heating rate of 20°C min^{-1} from temperature at -60°C to 200°C under a constant flow of nitrogen of 60 ml min^{-1}.

A universal testing machine (model 5567) from Instron Co., Ltd. was used to measure flexural strength and modulus of the blends according to ASTM D 790M-94. The test method used was a three-point bending mode with a support span of 48 mm at a constant cross head speed of 1.2 mm min^{-1}. The dimension was 25 mm in width, 60 mm in length, and 2 mm in thickness. The average value from three samples was reported.

Biodegradability of the PLA/PBS blends was carried out using specimen with 12.7 mm in length and 3.3 mm in width. The specimens were buried into the 20 mm depth of soil having pH of 7.43 and moisture of 65.99%. The samples were then weighed for every ca. 7 days for total period of 28 days.

3. Results and Discussion

3.1 Chemical structure of the PLA/PBS blends

The FTIR spectra of the PLA/PBS blend at 80/20 at various radiation doses is illustrated as shown in figure 1. It was noticed that changes in the absorption peak characteristics of PLA after electron beam radiation at 0-150 kGy was observed. The appearance of absorption peaks at 1757, 1088 and 2994 cm^{-1} assigned to the C=O stretching and C-O stretching of ester group and C-H stretching of the blends, respectively. These characteristic peaks were similar to as those of previous report [3]. Polymer chain scissoring and crosslinking mechanisms were believed to occur at the same time during electron beam radiation as seen from the changes in transmittance of ester linkage and C-H stretching. Moreover, a small shoulder at 1663 cm^{-1} assigned to C=C was noticed from the PLA/PBS blend indicating the C=C was generated after radiation.

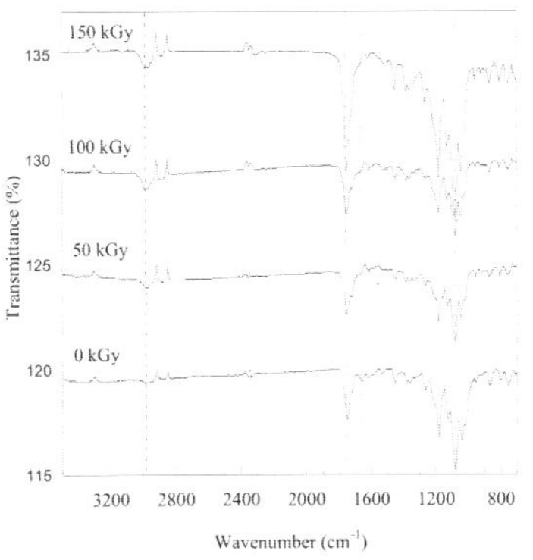

Figure 1. FT-IR Spectrum of PLA/PBS blend at weight ratio of 80/20 at various radiation doses.

3.2 T_m of the PLA/PBS blends

The T_m of the PLA/PBS blends at 100/0, 80/20 and 0/100 at various radiation doses were observed using DSC as seen in Table 1. It was noticed that the T_m of the pure PLA was 155°C which was higher than that of the pure PBS at 115°C. The results were in correspondence with those of previous report [4]. The value of PLA/PBS blend at 80/20 was 154°C which was slightly lower than that of the pure PLA. After irradiated, the T_m of the PLA/PBS blends tended to decrease with an increase radiation doses as a result of molecular crosslinking or molecular chain scissoring leading to a less order-packed of the polymer chain.

Table 1. T_m of the PLA/PBS blends at various radiation doses.

PLA/PBS at various weight ratios	T_m (°C)			
	0 kGy	50 kGy	100 kGy	150 kGy
100/0	155	154	151	149
80/20	154	153	150	148
0/100	115	111	111	111

3.3 Mechanical properties of the PLA/PBS blends

Flexural test of PLA/PBS blends was carried out to systematically investigate effects of radiation doses on mechanical properties of the blends as shown in figure 2.

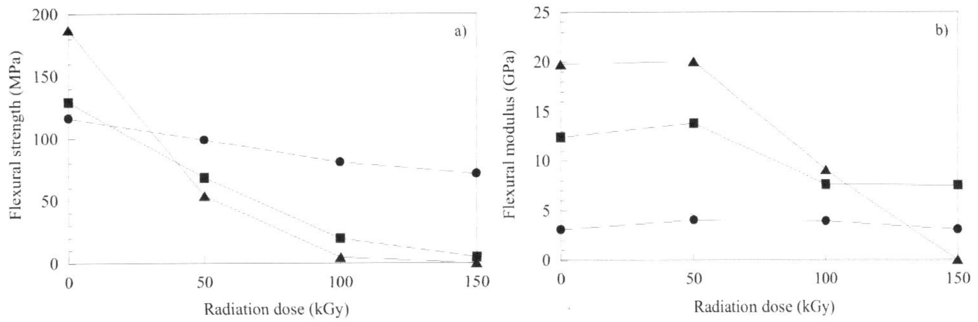

Figure 2. Mechanical properties of PLA/PBS blends at different blending weight ratios: (●) 100/0, (■) 80/20 and (▲) 0/100 after irradiated at various radiation doses: a) flexural strength and b) flexural modulus.

It was found from figure 2a) that an un-radiated PLA showed flexural strength value of 187 MPa which was higher than an un-radiated PBS having the value of 116 MPa while the strength of the PLA/PBS at weight ratio of 80/20 was 129 MPa. However, the values of the PLA/PBS blends at 100/0 and 80/20 significantly decreased after radiated at 50, 100 and 150 kGy. Interestingly, the PLA/PBS at 0/100 showed a slight reduction in the flexural strength value. A higher radiation dose resulted in a reduction in flexural strength of the PLA/PBS due to some degradation degree of the blends.

From figure 2b) it was observed that the flexural modulus of the specimens was improved with radiation dose at 50 kGy which might be due to the greatest degree of crosslinking. However, the modulus value of the samples decreased with increasing radiation doses. It was attributed to the degradation of the PLA/PBS blend after radiated.

3.4 Biodegradability of the PLA/PBS blends

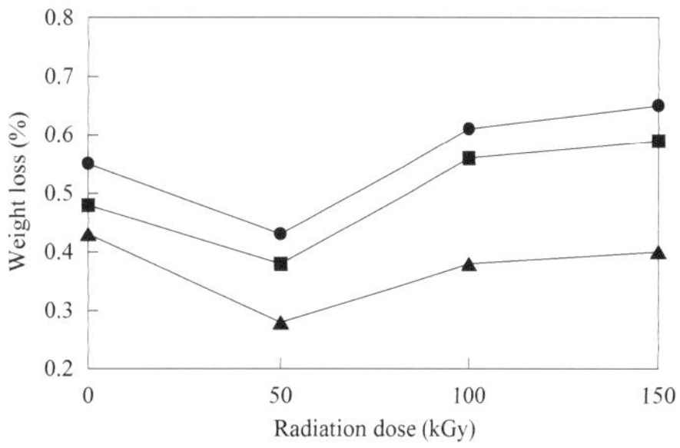

Figure 3. Weight loss of PLA/PBS blends after biodegradation at different weight ratios: (●) 100/0, (■) 80/20 and (▲) 0/100.

The biodegradation tests of the PLA/PBS at various weight ratios and different radiation doses were carried out for 28 days as illustrated in figure 3. In comparison at the same radiation doses, the PLA/PBS blend at 100/0 exhibited the greatest weight loss followed by the PLA/PBS blends at 80/20 and 0/100, respectively. After irradiated at 50 kGy, it was found that the weight loss of the specimen tended to decrease. At higher radiation doses than 50 kGy i.e. 100 and 150 kGy, the weight loss of the samples increased. The results were in accordance with the flexural modulus results i.e. the PLA/PBS after irradiated at 50 kGy showed the greatest modulus value among the others. It might because this radiation dose provided the highest degree of crosslinking while the degradation took place at radiation doses higher than 50 kGy.

4. Conclusions

The effects of the PLA/PBS blending weight ratios and radiation doses on chemical, mechanical, thermal and biological properties of the PLA/PBS blends were investigated in this research. The FTIR spectra indicated some structural change at different radiation doses. The melting temperatures of the PLA/PBS blends were decreased with increasing radiation doses. The results from flexural test revealed that the flexural strength of the specimens decreased with greater radiation doses. Interestingly, the PLA/PBS at 80/20 after radiated at 50 kGy showed the greatest flexural modulus value resulting in the lowest weight loss after biodegradation test.

5. References

[1] Dorgan J, H.L. and Mang M 2000 *J. Polym. Environ.* 8.
[2] Ji D, Xiaorong ZL, Wu F, Xie B and Yang M 2014 *J. Appl. Polym. Sci.* **131** 39580.
[3] Bureepukdee C, Suttiruengwong S and Seadan M 2015 *Mater. Sci. Eng.* **87** 012070.
[4] Saeed U, Nawaz MA and Al-Turaif HA 2018 *J. Compos. Mater.* **52** 2641.

Acknowledgments

This work is financially supported from the Research Promotion and Technology Transfer Center (RPTTC), Faculty of Liberal Arts and Science, Kasetsart University, Thailand.

IOP Conf. Series: Materials Science and Engineering **526** (2019) 012013 doi:10.1088/1757-899X/526/1/012013

Comparative neutron-shielding properties of metal oxide/HDPE composites using a Monte Carlo Code of PHITS

W Poltabtim[1], D Toyen[1] and K Saenboonruang[1,2,*]

[1] Department of Applied Radiation and Isotopes, Faculty of Science, Kasetsart University, Bangkok 10900, Thailand
[2] Specialized center of Rubber and Polymer Materials in agriculture and industry (RPM), Faculty of Science, Kasetsart University, Bangkok, 10900, Thailand

* Corresponding author: kiadtisak.s@ku.th, fscikssa@ku.ac.th

Abstract. Development of efficient neutron-shielding materials has become one of the most important issues in radiation safety. In order to estimate shielding performances of the materials of interest, simulations of neutron-shielding characteristics using well-proved methods are necessary. In this work, a Monte Carlo code of Particle and Heavy Ion Transport code System (PHITS) was used to simulate neutron-shielding properties of HDPE composites that were added with either B_2O_3, Sm_2O_3, Gd_2O_3, or CdO as a filler at various contents. The energy of the incoming neutrons was fixed at 0.025 eV with the beam direction pointing at the right angle to the material's surface. The results showed that Gd_2O_3/HDPE composites had the lowest neutron transmission ratio compared with other fillers at the same contents, implying the highest neutron-shielding ability. Furthermore, the simulations also showed that the values of neutron transmission ratios of all fillers decreased with increasing filler's contents. Other information such as half-value-layer (HVL) and tenth-value-layer (TVL) were also simulated and reported in this work. We encouraged further investigation of this work by comparing the simulated results with the actual experiment in order to validate the accuracy of the simulation.

1. Introduction

Neutron technology has been utilized widely nowadays including water film thickness measurement [1], Boron Neutron-Capture Therapy (BNCT) [2], and analysis of ancient artifacts [3]. However, despite the benefits of neutron utilization, excessive exposure to neutrons could fatally harm users, ranging from skin burns, vomiting, dizziness, and possible death [4, 5].

To prevent and/or to reduce risks from radiation exposure, sufficient neutron shielding is required. In principle, good neutron shielding materials would compose of a combination of light elements such as hydrogen (H) or helium (He) and high-neutron-absorption-cross-section (σ_{abs}) elements such as boron-10 (^{10}B). Examples of such materials are B_2O_3/paraffin composites [6], B_2O_3/natural rubber (NR) [7], and nano-B_4C/high density polyethylene (HDPE) [8]. In particular, HDPE ($(C_2H_4)_n$) is one of common materials used in neutron shielding applications, mainly due to its high hydrogen content, which could effectively thermalize, i.e. reduce energy of, fast neutrons with higher efficiencies than other polymers such as NR or low density polyethylene (LDPE) [9]. HDPE also offers exceptional strength to density ratio and is able to withstand high temperatures (120°C).

In addition to light elements, high-σ_{abs} elements, especially ^{10}B, are also essential in thermal neutron attenuation. These elements could interact with incoming thermal neutrons, creating nuclear reactions and releasing other types of ionizing radiations such as alpha ($^{4}He^{2+}$) or proton ($^{1}H^{+}$) particles that are simpler to being shielded. One of the most commonly used elements having high σ_{abs} is boron, of which ^{11}B (80% abundance) has σ_{abs} of 0.0055 barns, while ^{10}B (20% abundance) has σ_{abs} of 3835 barns, making ^{10}B a better neutron absorber than a major isotope, ^{11}B. However, using purified ^{10}B as a neutron absorber would require complicated procedures to extract ^{10}B from natural boron (^{nat}B), hence, it would be much simpler and more cost effective to use ^{nat}B as a neutron absorber, although the resulting σ_{abs} would be reduced to 767 barns. The nuclear reactions between ^{10}B nuclei and thermal neutrons are shown in Eq. 1 and Eq. 2, in which 94% of all nuclear reactions would produce 0.48-MeV gamma (γ) rays.

$$^{10}B^{5+} + n \rightarrow {}^{4}He^{2+} + {}^{7}Li^{3+}\ (6\%) \tag{1}$$
$$^{10}B^{5+} + n \rightarrow {}^{4}He^{2+} + {}^{7}Li^{3+} + \gamma\ (94\%) \tag{2}$$

However, ^{nat}B is not the only element that could be used as a neutron absorber, other elements such as samarium (^{nat}Sm), gadolinium (^{nat}Gd), and cadmium (^{nat}Cd) that have σ_{abs} of 5,922 barns, 49,700 barns, and 2,520 barns, respectively, also have potentials to be used as effective neutron absorbers due to their high σ_{abs}. The neutron reactions between ^{149}Sm, ^{157}Gd, or ^{113}Cd nuclei (having the highest σ_{abs} among their respective isotopes) and thermal neutrons are shown in Eq. 3–5.

$$^{149}Sm^{62+} + n \rightarrow {}^{4}He^{2+} + {}^{146}Nd^{60+} \tag{3}$$
$$^{157}Gd^{64+} + n \rightarrow {}^{158}Gd^{64+} + \gamma \tag{4}$$
$$^{113}Cd^{48+} + n \rightarrow {}^{114}Cd^{48+} + \gamma \tag{5}$$

As seen by the high σ_{abs} of ^{nat}B, ^{nat}Sm, ^{nat}Gd, and ^{nat}Cd, this work aimed to simulate their thermal neutron shielding properties of their respective oxide compounds (B_2O_3, Sm_2O_3, Gd_2O_3, and CdO) in HDPE using a Monte Carlo code of Particle and Heavy Ion Transport code System (PHITS) [10, 11]. Important properties including neutron transmission ratio (I/I_0), half-value-layer (HVL), and tenth-value-layer (TVL), were thoroughly determined using the results from this simulation.

2. Experimental

PHITS is a multi-purpose Monte Carlo simulation package that could be used in various applications including the design of spacecraft and accelerator facilities. In this work, PHITS was used to simulate thermal neutron shielding properties of metal oxide/HDPE composites by considering their transport and collision processes. To set up the simulation, a thermal neutron beam (0.025 eV) with 0.1-mm diameter was pointed at a right angle to the surface of shielding materials (z axis). The sizes of all samples were 40 cm × 40 cm with their thicknesses varied from 0.2–1.0 cm in 0.2-cm increments and the detector with 100% detection efficiency was located at 20 cm away from the surface of the shielding materials. The setup of the simulation is shown in Fig.1. It should be noted that all simulations were performed in air conditions.

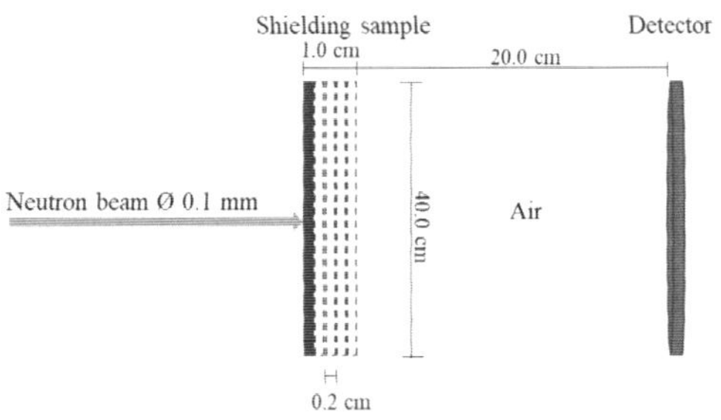

Fig. 1. Setup of the simulation to determine thermal neutron attenuation using PHITS.

Contents of all metal oxides in the HPDE composites used for the simulations were 10wt%, 20wt%, and 30wt%, respectively. The neutron attenuation could be calculated using Eq. 6;

$$I = I_0 e^{-\mu x} \tag{6}$$

where I, I_o, μ, and x are transmitted neutron intensity, initial neutron intensity, linear attenuation coefficient, and thickness of the sample. The values of HVL and TVL were determined using Eq. 7 and Eq. 8. It should be noted that μ used in the calculation of HVL and TVL was calculated using the results of the samples with 2-mm thickness due to the least effect from build-up factors [12].

$$HVL = \ln(2)/\mu \tag{7}$$
$$TVL = \ln(10)/\mu \tag{8}$$

3. Results and Discussion

As shown in Fig. 2, Gd_2O_3/HDPE composites could attenuate thermal neutrons with the highest efficiency as seen by the lowest numbers of transmitted neutrons. This was due to the fact that ^{nat}Gd nuclei have the highest value of σ_{abs}, leading to a larger probability of neutron interactions and absorption. Fig.3 illustrates effects of sample thicknesses to neutron shielding ability. The results showed that, as samples became thicker, less neutron could transmit. This was mainly due to higher numbers of metal oxides in thicker samples that could interact with neutrons, resulting in more neutron attenuation.

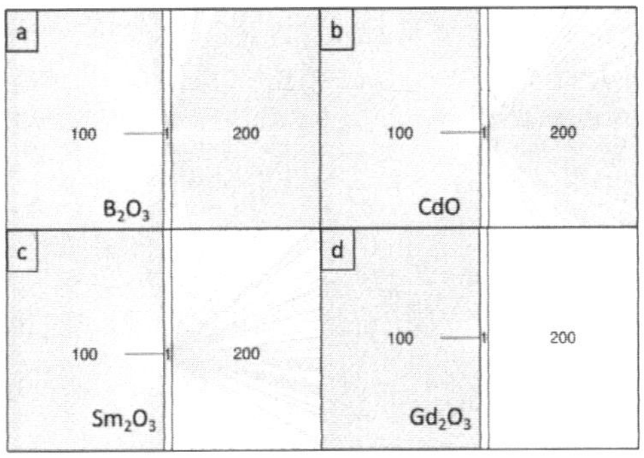

Fig.2. Results from PHITS simulation showing thermal neutron transmission of (a) B_2O_3/HDPE, (b) CdO/HDPE, (c) Sm_2O_3/HDPE, and (d) Gd_2O_3/HDPE composites. All samples were 10 mm thick with 30wt% metal oxide content.

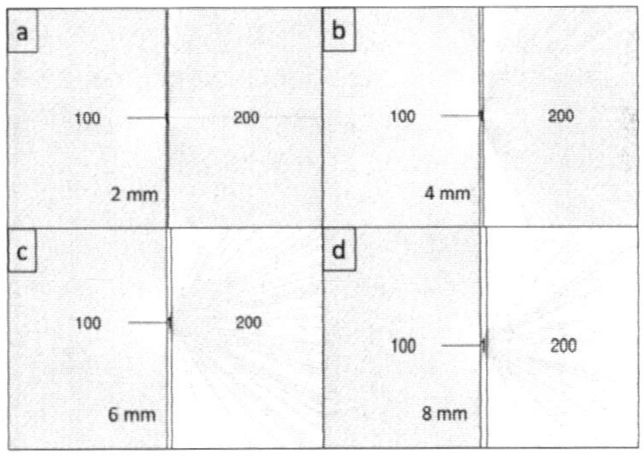

Fig.3. Results from PHITS simulation showing thermal neutron transmission of 10wt% Gd_2O_3/HDPE composites with (a) 2 mm, (b) 4 mm, (c) 6 mm, and (d) 8 mm thickness.

In terms of the values of neutron transmission ratios (I/I_0) shown in Fig. 4 and HVL/TVL values presented in table 1, it was obvious that Gd_2O_3/HDPE composites had the highest neutron shielding ability as illustrated by the lowest I/I_0, HVL, and TVL values compared with other composites. Another interesting result was that the ability to attenuate neutrons improved with increasing contents of metal oxide. This was because, as contents of metal oxide became larger, numbers of high-σ_{abs} nuclei that could effectively absorb thermal neutrons increased, leading to more interactions and less neutron transmission.

Table 1. Values of HVL and TVL of all HDPE composites.

Content (wt%)	Metal oxide							
	B_2O_3		Sm_2O_3		Gd_2O_3		CdO	
	HVL (cm)	TVL (cm)	HVL (cm)	TVL (cm)	HVL (cm)	TVL (cm)	HVL (cm)	TVL (cm)
10	6.23	20.71	3.30	10.97	0.44	1.47	3.24	10.76
20	2.66	8.84	1.62	5.41	0.23	0.78	1.73	5.75
30	1.73	5.77	1.11	3.70	0.17	0.58	1.27	4.23

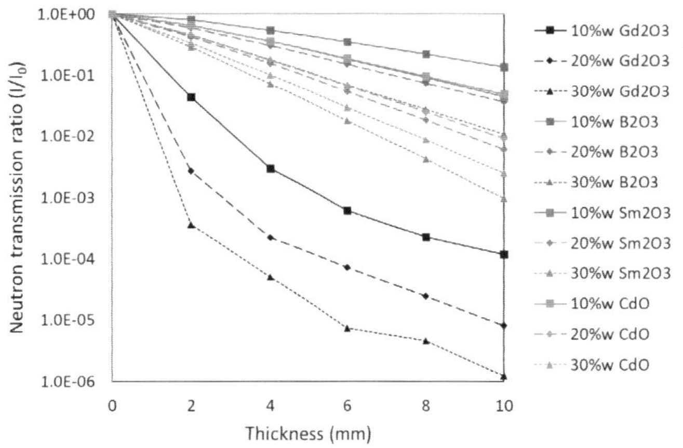

Fig.4. Neutron transmission ratios (I/I_0) of all HDPE composites. Note that the data are plotted in log scale.

4. Conclusions

PHITS was used to simulate thermal neutron shielding properties of HDPE composites that were added with either B_2O_3, Sm_2O_3, Gd_2O_3, or CdO as a neutron absorber. The results indicated that increasing metal oxide contents in HDPE composites led to the increases in thermal neutron shielding ability as seen by the decreases in neutron transmission ratios (I/I_0), HVL, and TVL. Furthermore, the results also showed that Gd_2O_3/HDPE composites had higher neutron shielding ability compared with other composites at the same contents.

5. References

[1] Dupont J, Mignot G, Zboray R and Prasser H M 2016 *J. Nucl. Sci. Technol.* **53** 673–81
[2] Chadha M, Capala J, Coderre J A, Elowitz E H, Iwai J, Joel D D, Lie H B, Wielopolski L and Chanana A D 1998 *Int. J. Radiat. Oncol. Biol. Phys.* **40** 829–34
[3] Boulanger M T, Buchanan B, O'Brien M J, Redmond B G, Glascock M D and Eren M I 2015 *J. Archaeol.* **53** 550–8
[4] Pinkerton L E, Waters M A, Hein M J, Zivkovich Z, Schubauer-Berigan M K and Grajewski B 2011 *Am. J. Ind. Med.* **55** 25–36
[5] El-Taher A and Uosif M A M 2006 *J. Phys. D: Appl. Phys.* **39** 20
[6] Toyen D and Saenboonruang K 2017 *J. Nucl. Sci. Techol.* **54** 871–7
[7] Ninyong K, Wimolmala E, Sombatsompop N and Saenboonruang K 2017 *Polym. Test.* **59** 336–43
[8] Kim J, Lee B C, Uhm Y R and Miller W 2014 *J. Nucl. Mater.* **453** 48–53
[9] Zuin L, Innocenti F, Fabris D, Lunardon M, Nebbia G, Viesti G, Cinausero M, Fioretto E, Prete G, Palomba M and Pantaleo A 2000 *Nucl. Instrum. Methods Phys. Res. A.* **449** 416–26
[10] Iwase H, Niita K and Nakamura T 2002 *J. Nucl. Sci. Techol.* **39** 1142–51
[11] Niita K, Sato T, Iwase H, Nose H, Nakashima H and Sihver L 2006 *Radiat. Meas.* **41** 1080–90
[12] Suteau C and Chiron M 2005 *Radiat. Prot. Dosimetry.* **116** 489–9

Acknowledgements

Authors would like to acknowledge technical supports from the Thailand Institute of Nuclear Technology (TINT) and Japan Atomic Energy Agency (JAEA). Authors also would like to thank the Specialized center of Rubber and Polymer Materials for agriculture and industry (RPM), Faculty of Science, Kasetsart University for the publication support.

IOP Conf. Series: Materials Science and Engineering **526** (2019) 012014 doi:10.1088/1757-899X/526/1/012014

Effects of Bromide-mixing in Methylammonium Lead Iodide on Photovoltaic Properties of ZnO-based Perovskite Solar Cells

C Khaywimut[1,2], C Bhoomanee[1,2], S Choopun[1,2,3], D Wongratanaphisan[1,2,3], A Pengpad[4], P Ruankham[1,2,3,*]

[1] Department of Physics and Material Science, Faculty of Science, Chiang Mai University, Chiang Mai, 50200, Thailand
[2] Thailand Center of Excellent in Physics (ThEP Center), Commission Higher Education, Bangkok 10400, Thailand
[3] Research Center in Physics and Astronomy, Faculty of Science, Chiang Mai University, Chiang Mai, 50200, Thailand
[4] Department of Physics, Faculty of Science, Khon Kaen University, Khon Kaen, 40002, Thailand

*Corresponding author: pipat.r@cmu.ac.th

Abstract. To improve efficiency of ZnO-based perovskite solar cells, a mixture iodide-bromide (I-Br) of $FA_{0.4}MA_{0.6}PbI_{3-2x}Br_{2x}$ perovskite structure was prepared by two-step sequential deposition method and used as photo absorber layer. Solar cells with structure of ITO/ZnO/PCBM/perovskite/P3HT/Au were fabricated and characterized. The perovskite films were prepared by mixing precursor PbI_2 and $PbBr_2$ at various molar ratio. It was found that the grain sizes and crystallite sizes of perovskite crystal were comparable for different Br doping. The existence of Br ions in perovskite layer was confirmed. In addition, the increase in band gap energy was observed. The highest average power conversion efficiency (PCE) of 2.17% was achieved from the device with $FA_{0.4}MA_{0.6}PbI_{2.8}Br_{0.2}$ layer. This PCE value is higher than the $FA_{0.4}MA_{0.6}PbI_3$ device which exhibited 1.53%-efficiency. The enhancement in PCE could be explained by the development of light absorption and charges transport.

1. Introduction

Organometal halide perovskite structured of $CH_3NH_3PbI_3$ ($MAPbI_3$) is widely used as light absorber material for solar cells due to its various benefits such as high absorption coefficient, ease of synthesis, applicability for flexible substrate, and non-rare earth elements [1-3]. However, their durability is still problematic [4]. To deal with the issue, bromide (Br) ions were added into the structure of $MAPbI_3$ [5] since they can improve the stability of solar cells, reduce charge recombination and enhance the efficiency [6-7].

In addition, the incorporation of formamidinium (FA^+) cation to methylammonium (MA^+) was another choice to develop the efficiency of solar cells because the existence of FA^+ ions can reduce the band gap

IOP Conf. Series: Materials Science and Engineering **526** (2019) 012014 doi:10.1088/1757-899X/526/1/012014

energy of light absorber layer, increase the crystallite size [8] and enhance carrier lifetime [9]. Chen *et al.* proved that the FA^+:MA^+ ratio of 4:6 provided the optimum PCE [9].

In this research the $FA_{0.4}MA_{0.6}PbI_{3-2x}Br_{2x}$ perovskite light absorber layer were prepared by two-step sequential deposition method. The $PbI_{(1-x)2}Br_{2x}$ layer were first deposited on the ZnO layer. Then, the perovskite light absorber layer was formed with the mixture of FAI and MAI solution. With the benefits of Br ions and FA^+ cation in the photo-absorber, the PCE of perovskite solar cells increased expectedly.

2. Experimental Details

A ZnO compact layer was deposited on a UV-treated indium doped tin oxide (ITO) substrate by spin coating a colloidal ZnO solution. This deposition method was patterned after the procedure reported by Horachit *et al.*[10]. Then PCBM was deposited on the ZnO layer by spin-coating a solution of 10 mg/ml PCBM in chlorobenzene (CB) in a N_2-filled glovebox. A $PbI_{2-2x}Br_{2x}$ precursor solution was prepared by mixing of PbI_2 and $PbBr_2$ in a mixture of 0.5 ml dimethylformamide (DMF) and 60μl of 4-*tert*-butylpyridine (tBP). The tBP was used as additive to form porous precursor layer [11]. The x values were varied as 0, 0.1 0.2 and 0.3. Then, the precursor solution was spin-coated at 3000 rpm for 30 s and the films were annealed at 100 C° for 1 hr in a N_2-filled glovebox. Next a mixture of 4.3 mg/ml of FAI and 6 mg/ml of MAI solution in isopropanol was dropped on the $PbI_{2-2x}Br_{2x}$ precursor layer and spun at 2000 rpm for 30s. films were then annealed at 75 C° for 20 min. A hole transporting layer (HTL) was prepared by spin coating a mixture of 20 mg/ml P3HT, 8 μl of 280 mg/ml Co (II) in acetonitrile, 8 μl of 4-tBP and 16 μl of acetonitrile in 1 ml of CB at 1000 rpm for 30 s. Lastly Au metal electrode was coated on HTL layer for the electrode via thermal evaporation under 10^{-4} bar vacuum condition.

The morphology of perovskite layer was observed by field emission scanning electron microscope (FE-SEM, JEOL JSM-6335F). Chemical composition of perovskite layer was analysed by energy-dispersive x-ray spectroscopy (EDS). Crystallite properties was investigated by X-ray diffraction (XRD) analysis. Optical properties of perovskite layer were studies via UV-visible spectroscopy and band gap energy (E_g) of perovskite layer was calculated by Tac's equation [12]. Photovoltaic performance was measured under AM 1.5 condition with active area are about 38 mm^2.

3. Results and Discussion

The FE-SEM images of the $FA_{0.4}MA_{0.6}PbI_{3-2x}Br_{2x}$ layer with different x content values were showed in Figure 1. It was found that the average grain sizes of these materials were comparable (241, 249, 247 and 237 nm for x= 0, 0.1, 0.2, and 0.3 respectively).

The element compositions analyzed by EDS for $FA_{0.4}MA_{0.6}PbI_{3-2x}Br_{2x.}$ layer were summarized in Table 1. These results confirmed the presence of Br ions in the perovskite layer. It is seen that the atomic ratios of Br/Pb calculated from the detected data were comparable to the ratio of the initially mixed precursor. On the other hand, the ratio of I/Pb from the detected data were larger than the expected value. This maybe occurs due to the PbI_2 residue in the film.

Figure 1. Top view FE-SEM images of $FA_{0.4}MA_{0.6}PbI_{3-2x}Br_{2x}$ layer for different x content value.

Table 1. Element compositions, atomic ratio and crystallitte size of $FA_{0.4}MA_{0.6}PbI_{3-2x}Br_{2x}$ layers.

x	Structure	Atomic %			Theory		Experiment		Crystallite size (nm)
		Pb	I	Br	Br/Pb	I/Pb	Br/Pb	I/Pb	
0	$FA_{0.4}MA_{0.6}PbI_3$	21.59	78.41	-	-	3	-	3.63	32
0.1	$FA_{0.4}MA_{0.6}PbI_{2.8}Br_{0.2}$	20.17	73.64	6.19	0.2	2.8	0.31	3.65	34
0.2	$FA_{0.4}MA_{0.6}PbI_{2.6}Br_{0.4}$	23.02	67.82	9.16	0.4	2.6	0.40	2.95	34
0.3	$FA_{0.4}MA_{0.6}PbI_{2.4}Br_{0.6}$	23.33	62.03	14.64	0.6	2.4	0.63	2.66	32

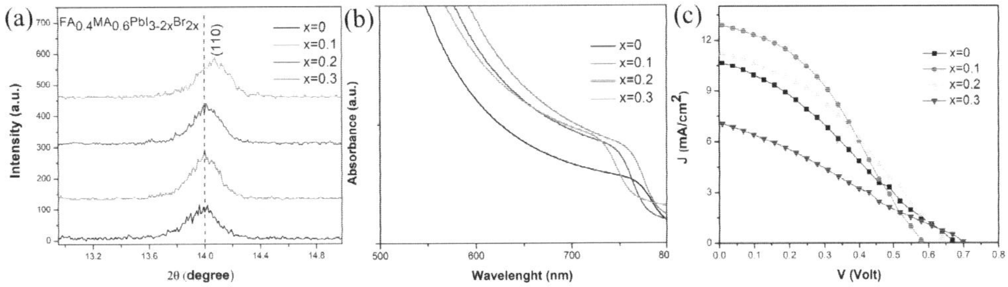

Figure 2. (a) XRD peaks at (110) plane and (b) absorption spectra of $FA_{0.4}MA_{0.6}PbI_{3-2x}Br_{2x}$ layers. (c) shows *J-V* curves of perovskite solar cells with using $FA_{0.4}MA_{0.6}PbI_{3-2x}Br_{2x}$.

As confirmed by XRD analysis, there were both tetragonal perovskite phase (XRD peaks at 14°, 28.2° and 31.6°) and residue PbI_2 phase (XRD peak at 12.6°) in the films. Figure 2a showed the diffraction peak at approximately 14° that refers to (110) plane [13] of the perovskite structure. It was seen that, when the x value increased, the diffraction peak shifted from lower peak position to a larger value. By using Bragg's equation [14, 15], it was found that the *d*-spacing and lattice constant decreased from 6.33 Å to 6.29 Å and 8.95 Å to 8.89 Å with increasing x content from 0 to 0.3. These phenomena can be explained by the substitutional of Br ions to I ions in the perovskite structure. In addition, calculated crystallite sizes of these materials (in Table 1) were comparable. This investigation is consistent with the FE SEM observation, implying that the Br doping does not change the grain size and crystallite size of perovskite materials.

The absorption spectra of $FA_{0.4}MA_{0.6}PbI_{3-2x}Br_{2x}$ perovskite layers were in showed Figure 2(b). The increase in E_g (from 1.56 eV for x = 0 to 1.62 eV for x = 0.3) was found. This increment can be ascribed to the Br doping. Figure 2c showed current density-voltage (*J-V*) characteristic curves of perovskite solar cells using $FA_{0.4}MA_{0.6}PbI_{3-2x}Br_{2x}$ as phot-absorber. The average photovoltaic parameters were summarized in table 2. It was found that the highest average PCE of 2.17% was obtained from the device with x= 0.1. In addition, open circuit voltage (V_{OC}) for the devices with x = 0, 0.1 and 0.2 were comparable but the V_{OC} value of the one with x = 0.3 was higher. The increase in V_{OC} is attributed to the enlargement of E_g when more Br atoms incorporate into the crystal as confirmed by the calculated E_g and the results reported in reference [16].

Short-circuit current density (J_{sc}) of the fabricated solar cells was relatively high for x= 0.1 and 0.2. The higher J_{sc} can be explained by two reasons. First, as seen in Figure 2b, the absorbance of the perovskite layer with x = 0.1 provided the strongest absorption so the perovskite layer has ability to generate more excitons than the other films. Finally, the series resistance (R_s), which is commonly related to charge transportation [17], of the devices with x = 0.1 showed the lowest value, therefore, better charge transport could be occurred

Table 2. The average photovoltaic parameters (over 6 cells prepaed at the same condition) at different x content value.

x	PCE (%)	V_{oc} (V)	J_{sc} (mA/cm^2)	FF	R_s (Ωcm^2)
0	1.53 ± 0.33	0.61 ± 0.06	8.60 ± 1.67	0.30 ± 0.02	54.38 ± 10
0.1	2.17 ± 0.46	0.62 ± 0.02	9.95 ± 2.02	0.35 ± 0.03	36.27 ± 6.0
0.2	1.88 ± 0.44	0.61 ± 0.06	9.99 ± 1.32	0.31 ± 0.05	45.77 ± 25
0.3	1.39 ± 0.13	0.66 ± 0.03	6.94 ± 0.63	0.31 ± 0.03	40.62 ± 10

4. Conclusion

The $FA_{0.4}MA_{0.6}PbI_{3-2x}Br_{2x}$ perovskite solar cells based on ZnO ETL were successfully prepared via two-step sequential method. Br atoms were incorporated into the perovskite layer by using the mixture of PbI_2 and $PbBr_2$ as the precursors. The highest average PCE of 2.17 % was obtained from the device with x = 0.1. The improvement in PCE was due to the higher absorption, and the reduction of R_s

5. References

[1] Qin F, Meng W, Fan J, Ge C, Luo B, Ge R, Hu L, Jiang F, Liu T, Jiang Y, and Zhou Y 2017 *ACS Appl. Mater. Interfaces* **9** 26045-51
[2] Bahtiar A, Rahmanita S, and Inayatie Y D 2017 *IOP Conf. Ser.: Mater. Sci. Eng.* **196** 012037
[3] Lund P D, Halme J, Hashmi G, Asghar I, and Miettunen K 2018 *Flex. Print. Electron.* **3** 013002
[4] Popoola I K, Gondal M A, and Qahtan T F 2018 *Renew. Sus. Energ. Rev.* **82** 3127-51
[5] Tombe S, Adam G, Heilbrunner H, Apaydin D H, Ulbricht C, Sariciftci N S, Arendse C J, Iwuoha E, and Scharber M C 2017 *J. Phys. Chem. C* **5** 1714-23
[6] Li F, and Liu M 2017 *J. Mater. Chem. A* **5** 15447-59
[7] Gil-Escrig L, Miquel-Sempere A, Sessolo M, and Bolink H J 2015 *J. Phys. Chem. Lett.* **6** 3743-8
[8] Liu J, Shirai Y, Yang X, Yue Y, Chen W, Wu Y, Islam A, and Han L 2015 *Adv.Mater.* **27** 4918-23
[9] Chen J, Xu J, Xiao L, Zhang B, Dai S, and Yao J 2017 *ACS Appl. Mater. Interfaces* **9**, 2449−2458
[10] Horachit C, Intaniwet A, Choopun S, and Ruankham P 2018 *Opt. Quant. Electron.* **50** 379
[11] Shi Y, Wang X, Zhang H, Li B, Lu H, Ma T, and Hao C 2015 *J. Mater .Chem. A* **3** 22191-22198
[12] Ghobadi N 2013 *Int. Nano Lett.* **3** 2
[13] Park B-w, Philippe B, Jain S M, Zhang X, Edvinsson T, Rensmo H, Zietz B, and Boschloo G 2015 *J. Mater. Chem. A* **3** 21760-71
[14] Elton L R B, and Jackson D F 1966 *Am. J. Phys.* **34** 1036-8
[15] Tsai D S, Chin T S, Hsu S E, and Hung M P 1989 *Mater. Trans.* **30** 474-9
[16] Haque M A, Sheikh A D, Guan X, and Wu T 2017 *Adv. Energy Mater.* **7** 1602803
[17] Li Y, Ding B, Chu Q Q, Yang G J, Wang M, Li C X, and Li C J 2017 *Sci. Rep.* **7** 46141

Acknowledgement

This research was financially supported by the Development and Promotion of Science and Technology Talents Project (DPST) (Research Fund for DPST Graduate with First Placement No. 25/2557). This research was partially supported by Chiang Mai University and Thailand Center of Excellent in Physics (ThEP Center). The authors deeply grateful to R Panyatip, and K Chaisan for process assistance.

IOP Conf. Series: Materials Science and Engineering **526** (2019) 012015 doi:10.1088/1757-899X/526/1/012015

Manufacturing process and properties of lead-free natural rubber sponge for use in X-ray and gamma ray shielding applications

P Lim-aroon[1], E Wimolmala[1,*], N Sombatsompop[1] and K Saenboonruang[2,3,*]

[1] Polymer Processing and Flow (P-PROF) Research Group, Division of Materials Technology, School of Energy, Environment and Materials, King Mongkut's University of Technology Thonburi (KMUTT), Bangkok 10140, Thailand.
[2] Department of Applied Radiation and Isotopes, Faculty of Science, Kasetsart University, Bangkok 10900, Thailand.
[3] Specialized center of Rubber and Polymer Materials in agriculture and industry (RPM), Faculty of Science, Kasetsart University, Bangkok, 10900, Thailand.

*Corresponding author (E Wimolmala: ekachai.wim@kmutt.ac.th and K Saenboonruang: kiadtisak.s@ku.th, fscikssa@ku.ac.th)

Abstract. In this work, light-weighted, flexible, and lead-free X-ray/gamma-ray shielding natural rubber (NR) sponges were developed with the help of a blowing agent, namely Oxybis (benzene sulfonyl) hydrazide (OBSH) (its contents varied from 0, 8 to 16 parts per hundred parts of rubber; phr), and an X-ray/gamma protective filler, namely bismuth oxide (Bi_2O_3) (its content varied from 0, 100, 300, to 500 phr) in natural rubber (NR) composites. The results suggested that the densities and overall mechanical properties decreased with increasing contents of OBSH, while the increases in B_2O_3 resulted in the increases in tensile strength and hardness (Shore A), however, initially increases tensile strength and elongation at break but later decreases at 300 and 500 phr. In terms of radiation shielding properties, it was found that the ability to attenuate X-rays and gamma rays improved with increasing Bi_2O_3 contents. Hence, the overall properties investigated in this work suggested that the developed NR sponges could be used to attenuate X-rays and gamma rays efficiently with additional properties of being light-weighted and highly flexible, which are crucial for safety of radiation-related personnel and users.

1. Introduction

X-ray and gamma rays are two types of relatively high-energy electromagnetic waves that currently play important roles in various applications including cancer therapy [1], material imaging [2], and medical sterilization [3]. Despite the mentioned benefits, risks of exposure to both X-ray and gamma rays could potentially lead to fatal health hazards in radiation workers and users [4]. In order to prevent excessive X-ray and gamma exposure, sufficient shielding materials that could attenuate incoming intensities to safe levels must be introduced and used in all radiation-related workplaces. In principle, materials composing of high-atomic-number (Z) and high-density elements are normally used as shielding materials in order to prevent such risks. This is because high-Z materials such as lead (Pb) had relatively high probability of interactions with incoming X-ray and gamma rays, leading to higher energy transfer and larger attenuations [5].

However, Pb-containing materials could inversely pose environmental and health risks to product users as Pb is hazardous and lacks chemical stability [6]. As a result, Pb-free X-ray/gamma ray shielding materials have drawn special attention from researchers. In order to achieve similar shielding properties as Pb-containing materials, other high-Z and high-density elements such as bismuth (Bi) could also act as effective radiation protective fillers. In particular, for applications that require flexibility with good mechanical properties, natural rubber (NR) and synthetic rubbers such as EPDM could be used as the main matrices with the addition of Bi compound [7, 8]. Furthermore, in order to reduce weight/density of the materials, blowing agents could be added to the composites [9]. As a result, this work developed NR composites with the addition of bismuth oxide (Bi_2O_3) and Oxybis (benzene sulfonyl) hydrazide (OBSH), of which X-ray and gamma ray shielding, density, mechanical, and morphological properties of the composites were thoroughly investigated and reported.

2. Experimental

2.1 Materials and chemicals

Natural rubber (grade STR 5L) was used as the main matrix for this work. The vulcanizing formulations and their roles are given in Table 1.

Table 1. Material formulations of NR composites including chemicals and their roles.

Ingredient	**Content** (phr*)	**Function**
Natural rubber (STR 5L)	100 (part)	Matrix
ZnO	5.0	Activator
Stearic acid	3.0	Activator
Mercaptobenzothiazole; MBT	1.0	Accelerator
6PPD	1.0	Anti-oxidant
Silica	40.0	Reinforcement
Polyethylene glycol; PEG	5wt% added of silica	Activator
Paraffinic oil	10.0	Processing aid
Sulphur	2.0	Crosslinker
Bismuth oxide; Bi_2O_3	0, 100, 300, 500	X-ray & gamma protective filler
Oxybis (benzene sulfonyl) hydrazide; OBSH	0, 8 and 16	Blowing agent

** phr: parts per hundred parts of rubber by weight*

2.2 Preparation of NR composites

The NR was masticated on a laboratory two-roll mill (Yong Fong Machinery Co., Ltd., Thailand) for 3 min. The masticated NR was then compounded with the chemicals given in Table 1 using the two-roll mill for further 17 min. To prepare the NR sheets, hot compression molding at the temperature of 165°C with a hydraulic pressure at 170 kg/cm^2 was used for every individual compound.

2.3 Mechanical properties

The tensile properties including tensile modulus at 100% elongation, tensile strength, and elongation at break of the NR composites were investigated using a universal testing machine (Autograph AG-I 5kN, Japan) following ASTM D412-06 standard testing. For hardness measurement, all the samples were tested using a hardness durometer, Shore A, (Teclock GS-719G, Japan) according to ASTM D2240-05.

2.4 Density and morphology studies

Densities of the all NR sponges were determined using Archimedes water replacement technique according to ASTM D792-00. For the study of morphology and the distribution of fillers, the samples were investigated using a scanning electron microscope (SEM), JEOL (JSM-6610LV), at 10-kV accelerating voltage with back-scattered electrons (BSE). It should be noted that the samples were coated with gold using an ion sputtering device for 60 sec before performing SEM.

2.5 X-ray and gamma ray shielding properties

The X-ray shielding properties were tested using a 100-kV X-ray generator, which produced an X-ray beam with the Half-Value Layer (HVL) of 1.1 mmCu. The results of the X-ray attenuation were reported as equivalent to standard Pb (mmPb) and percentage of attenuation. For gamma ray shielding properties, the tests were performed using a ^{60}Co point source, which emits gamma rays with the average energy of

1.25 MeV. Linear attenuation coefficient (μ), percentage of gamma transmission, and HVL were reported for all samples.

3. Results and discussion

3.1 Mechanical properties of NR sponges

Table 2 Mechanical properties and densities of NR sponges with/without OBSH and Bi_2O_3. The values are presented as mean ± standard deviations of the mean.

Properties	NR composites											
	No OBSH / Bi_2O_3 (phr)				8-phr OBSH / Bi_2O_3 (phr)				16-phr OBSH / Bi_2O_3 (phr)			
	0	**100**	**300**	**500**	**0**	**100**	**300**	**500**	**0**	**100**	**300**	**500**
Tensile modulus (MPa)	1.3±0.1	2.9±0.1	3.6±0.2	3.3±0.2	1.2±0.1	1.7±0	1.7±0.2	0.9±0.1	1.0±0.1	1.4±0	1.2±0.1	1.5±0.1
Tensile strength (MPa)	10.5±1.2	15.6±0.9	10.4±1.1	7.6±0.8	4.6±0.3	8.3±1.1	5.0±0.6	3.4±0.3	3.9±0.3	4.8±0.7	3.3±0.2	2.3±0.3
% Elongation at break	489±19	396±14	296±30	250±17	289±12	324±21	249±16	385±28	306±14	255±16	227±10	168±11
Hardness (shore A)	46±1	60±1	69±1	71±1	22±1	34±1	35±1	39±2	19±2	27±2	31±3	48±3
Density (g/cm^3)	1.09	1.65	2.52	3.07	0.73	1.17	1.79	2.34	0.59	1.05	1.19	2.18

As shown in Table 2, it was found that, at the same Bi_2O_3 content, tensile properties and hardness (shore A) decreased with increasing OBSH content. This was due to OBSH produced small closed-cell structures inside NR matrix as shown in Fig. 1, reducing the ability to resist tensile forces from the presence of gas phases. On the other hand, at the same OBSH content, increasing Bi_2O_3 contents led to the increases in tensile modulus and hardness (shore A), probably due to the high rigidity of Bi_2O_3, while tensile strength and elongation at break initially increased when its content increased to 100 phr but later decreases as more Bi_2O_3 was added to the composites. This was due to the fact that Bi_2O_3, which is a metal oxide that could act as a co-activator, helped in chemical crosslinking, leading to higher crosslink density and higher values of tensile strength and elongation at break. However, as more Bi_2O_3 was added, filler-filler interactions occurred, leading to the agglomeration of Bi_2O_3 and resulting in worse particle dispersion and lower tensile strength/elongation at break [8].

In terms of the densities of NR composites, the values decreased with increasing OBSH contents due to small closed-cell structures that were produced from the addition of OBSH as shown in Fig.1. Furthermore, it was found that the densities of NR composites increased with increasing contents of Bi_2O_3, mainly due to the much higher density of Bi_2O_3 particles (density of Bi_2O_3 is 8.9 g/cm^3).

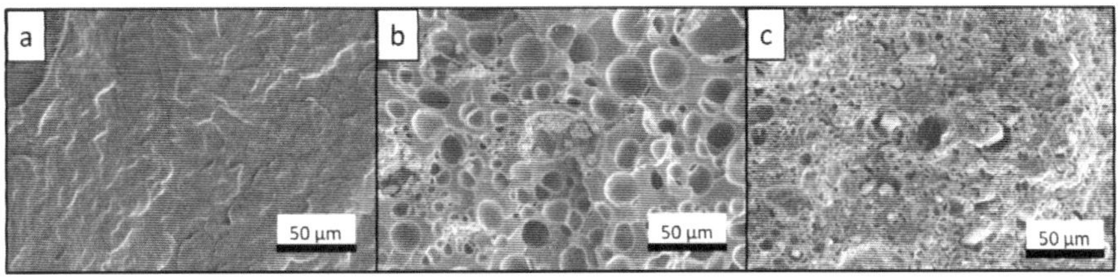

Fig. 1. Cross-sectional morphology of (a) NR composites, (b) NR composites with 8-phr OBSH, and (c) NR composites with 8-phr OBSH and 500-phr Bi_2O_3.

3.2 Gamma and X-ray shielding properties

Gamma ray and X-ray shielding properties of both NR solids and sponges with varying contents of Bi_2O_3 and OBSH are shown in Fig. 2 and Table 3, respectively. The results showed that increasing contents of Bi_2O_3 improved radiation shielding properties as seen by the increase in μ and the decreases in % gamma transmission and HVL. The improvement in gamma ray/X-ray shielding properties was mainly due to the increase of Bi elements in the NR composites that could interact and attenuate gamma rays/X-rays. However, the addition of OBSH led to the decrease in gamma shielding properties due to empty spaces in the NR matrix diluted effects of Bi element. Nonetheless, the values of linear attenuation coefficient at 500-phr Bi_2O_3 were in an acceptable range as they were similar to the values reported in ref. [7] and ref. [8], however, with smaller density and weight.

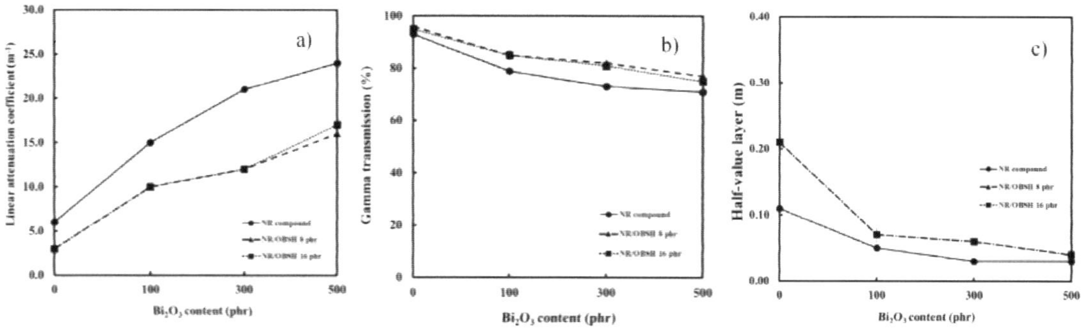

Fig. 2. Results of (a) linear attenuation coefficient, (b) gamma transmission, and (c) HVL of NR and 8-phr OBSH/NR composites with varying contents of Bi_2O_3.

Table 3. X-ray shielding properties of NR sponges

	8-phr OBSH/NR sponges	
X-ray shielding properties	**0-phr Bi_2O_3**	**500-phr Bi_2O_3**
Equivalent to Pb (mmPb)	0.01	0.34
% Attenuation	4.1	75.0

4. Conclusion

The results indicated that the addition of the blowing agent, OBSH, led to the decreases in overall mechanical, density, and gamma/X-ray shielding properties of NR sponges. Furthermore, the addition of Bi_2O_3 led to the increases in tensile modulus, hardness (shore A), density, and gamma ray/X-ray shielding properties, while initially increased tensile strength and elongation at break at the Bi_2O_3 content up to 100 phr but later decreased as more Bi_2O_3 was added to the NR composites. The overall results suggested that NR sponges with the addition of OBSH and Bi_2O_3 could be used as light-weighted, lead-free, and flexible gamma ray/X-ray shielding materials.

industry (RPM), Faculty of Science, Kasetsart University for the publication support.

5. References

[1] Krag D N, Weaver D L, Alex J C and Fairbank J T 1993 *Surg. Oncol.* **2(6)** 335-9
[2] Cloetens P, Barrett R, Baruchel J, Guigay J P and Schlenker M 1996 *J. Phys. D: Appl. Phys.* **29** 133
[3] Buchalla R, Schuttler C and Bogl K W 1995 *Radiat. Phys. Chem.* **46** 579-85
[4] Ghissassi F E, Baan R, Straif K, et al 2009 *Lancet Oncol.* **10** 751-2
[5] McCaffrey J P, Shen H, Downton B and Mainegra-Hing E 2007 *Med. Phys.* **34** 530-7
[6] Seregin I V and Ivanov V B 2001 *Russ. J. Plant Physiol.* **46** 1723-31
[7] Toyen D, Rittirong A, Poltabtim W and Saenboonruang K 2018 *Iranian Polym. J.* **27** 33-41
[8] Poltabtim W, Wimolmala E and Saenboonruang K 2018 *Radiat. Phys. Chem.* **153** 1-9
[9] Vahidifar A, Khorasani S N, Park C B, Naguib H E and Khonakdar H A 2016 *Ind. Eng. Chem. Res.* **55** 2407-16

Acknowledgement

Authors would like to acknowledge financial supports from Thailand Toray Science Foundation and the Office of the Higher Education Commission (OHEC) under the National Research University (NRU) program. Authors also would like to thank the Thailand Institute of Nuclear Technology (TINT) for the technical support and the Specialized center of Rubber and Polymer Materials for agriculture and

IOP Conf. Series: Materials Science and Engineering **526** (2019) 012016 doi:10.1088/1757-899X/526/1/012016

Electrical interaction between nanoparticle and surface of material

C. Summueang[1], S. Boonchui[1,*]
[1]Department of Physics, Kasetsart University, Bangkok, 10900, Thailand

*Corresponding author e-mail: fscistb@ku.ac.th

Abstract. We identify the interaction between nanoparticle and surface of material in the presence of an electric field. For theoretical calculation, we use the dyadic green function that includes with the method of image dipoles to analyze electrical interaction. We find that the electric field and the interaction potential which depend on the dielectric constant and surface roughness of material. Our numerical results demonstrate the electric field and the interaction potential which corresponds to the Van der Waals interaction and can be used to determine charges distribution on nanoparticle and surface of material.

1. Introduction

Particles in nanometer-scale were Influenced in the presence of an electric field that attracts on surface of metallic material and caused an impact of the electrical problem between nanoparticle and surface material that was acquired the interaction when nanoparticle is located close metallic substrate [1]. The interaction is related to electrical dipole moment and corresponds to the Van der Waals interaction [2,3].

We shall concentrate to our theoretical attention on the problem of nanoparticle in the colloidal state model and the interaction between a nanoparticle and the surface of a solid combined with the theoretical framework based on the physical properties of the polarization and magnetic Casimir-Polder interaction [4].

Particle colloidal behaviour is then compared to that of both isotropic and anisotropic minerals in order to gain further information regarding the surface charge distribution [5]. We have shown that it possible to explain the contamination of nanoparticle on the metallic substrate by using the dyadic Green's function [6] include with the method of images approach. By replacing the molecular particle–surface interaction with a molecular–image interaction, the image molecule due to the surface roughness [7] and the dielectric constant of the substrate. A magnitude of image molecule is proportional to the surface roughness and dielectric constant. This well-established technique can account for the force of attraction between a charge and an uncharged dielectric medium, Van der Waals interactions between molecules and surfaces. [8]

In this work, the main goal of our work is to analyze with numerical calculation the interaction potential from the consideration of the effect from dielectric constant to display through electric field graphs and predict charges distribution on particle and metallic material by comparing with the theoretical results. To compare with experimental results, we regard polarization of surface (surface roughness) and demonstrate our results through the interaction potential that are corresponding to Van der Waals interaction.

2. Theoretical calculation

We will present a theoretical approach for calculating the interaction between a finite-size nanoparticle of an arbitrary shape and the surface of a solid. Let us consider a nanoparticle located close to the metallic surface. The particle is situated in the medium with the dielectric constant ε_m that fills the half space $z > 0$, the surface with dielectric constant ε_s. The metallic surface is XY plane. The center of the particle is located at the point $\vec{R}_p = (0,0,d)$. To analyze the interaction, the potential in the system can be presented as [1]

$$U(d) = -\int_{V_p} dR_p P_i E_i - U(d \to \infty) \tag{1}$$

In order to find the electric polarization and electric field, one can write the free energy of the system that includes with the internal energy density and the energy of the polarization density in the following form [4]

$$E_f = \int_{V_p} d^3\vec{R}_p \, (U_{\text{int}} - \vec{P}\cdot\vec{E}) + \int_{R^3} d^3\vec{R}_p \, W \tag{2}$$

The electric dipole moment corresponding to the ground state of the system can be found from

$$\frac{\partial E_f(P_i)}{\partial P_i} = 0 \tag{3}$$

As a result of Eq. (3), one obtains the following connection between the dipole moment at the particle and the electric field

$$P_i(\vec{R}) \approx X_{ij}^{(p)}(\vec{R}_p)E_j^{(0)}(\vec{R}) + A_{ijkl}^{(p)}(\vec{R}_p)E_j^{(0)}(\vec{R})E_k^{(0)}(\vec{R})E_l^{(0)}(\vec{R}) \tag{4}$$

To calculate the electric field with the Lippmann-Schwinger equation by using dyadic Green's function method [9,10,11]

$$E_i(P_j) = E_i^{(0)} + \frac{k_0^2}{\varepsilon_0} \int_{V_p} dR_p G_{ij}(R_p, \omega) \cdot P_j(R_p) \tag{5}$$

The dyadic Green's function that is used in Eq. (5) is divided into the direct and the indirect term in the following form [10,12]

$$G_{ij}(\mathbf{R}, \mathbf{R}', \omega) = G_{ij}^{(D)}(\mathbf{R}, \mathbf{R}', \omega) + G_{ij}^{(I)}(\mathbf{R}, \mathbf{R}', \omega) \tag{6}$$

In the surface of metallic material take into account the surface roughness. Thus, One can put the effect of polarization from the surface in Eq. (2) in the form [7]

$$\vec{P}_s(\vec{r}) = \vec{p}b(\vec{r}) \tag{7}$$

and

$$\vec{p} = \gamma_z \hat{z}\hat{z} \cdot \vec{E}_i + \gamma_t (\hat{x}\hat{x} + \hat{y}\hat{y}) \cdot \vec{E}_i \tag{8}$$

IOP Conf. Series: Materials Science and Engineering **526** (2019) 012016 doi:10.1088/1757-899X/526/1/012016

3. Result and Discussion

3.1. Electric field of the system without surface roughness consideration

In order to obtain the numerical calculation, let us consider nanoparticle as a square particle in the presence of an applied external electric field in Z direction that came close to the surface of metallic material. To verify the theoretical calculation that corresponded with theoretical results without surface roughness consideration, in Figs. 1-3 was shown electric field by contour plot with each substrate in different the dielectric constant. One can see that the dielectric constant increases with the increase of the image particle that rise the interaction between particle and substrate increases. Consequently, the interaction depends on the dielectric constant of the substrate [4].

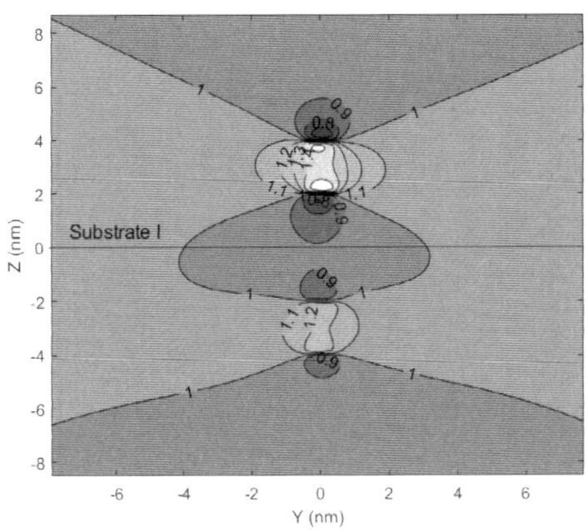

Figure 1. Electric field in the system from substrate I with $\varepsilon_s = 28$.

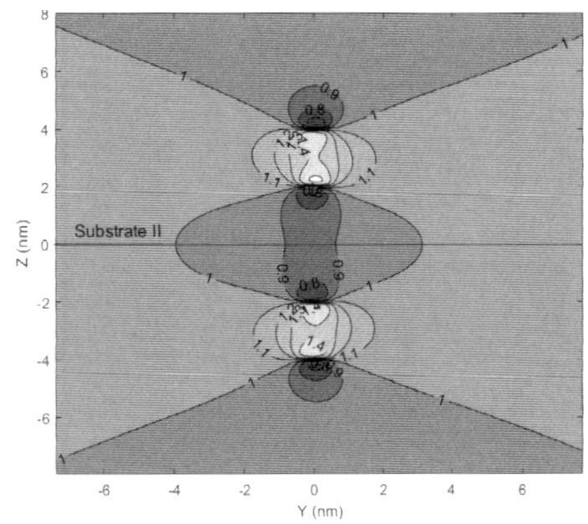

Figure 2. Electric field in the system from substrate II with $\varepsilon_s = 57$.

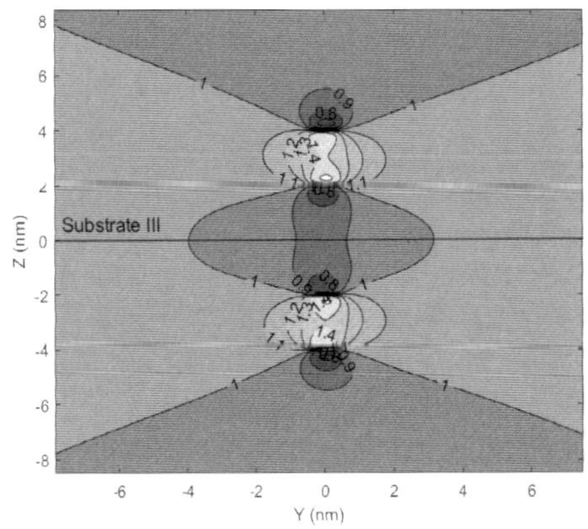

Figure 3. Electric field in the system from substrate III with $\varepsilon_s = 130$.

3.2. The experimental result and the interaction potential

The experimental result is shown in Fig.4, this is the probability of contamination of nanoparticle on the substrate. Due to the fact, that should be increased linearly according to the above mentioned. Thus, Let us consider the case when a nanoparticle interacts with a metallic substrate in the presence of surface roughness consideration and determine the interaction potential. The results of this calculation, are shown in Fig. 5. Obtained results are concentrated in substrate II that correspond to the experimental result. The potential depends on the dielectric constant and surface roughness and is not linear.

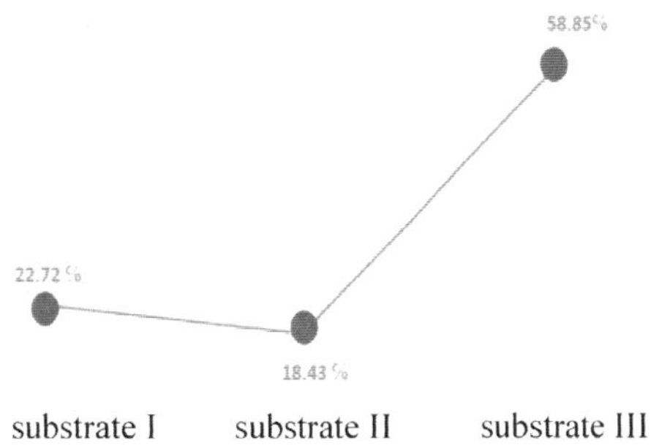

Figure 4. Experimental result illustrates Probability of contamination of nanoparticle on substrate (substrate I; $\varepsilon_s = 28$, substrate II; $\varepsilon_s = 57$, substrate III; $\varepsilon_s = 130$)

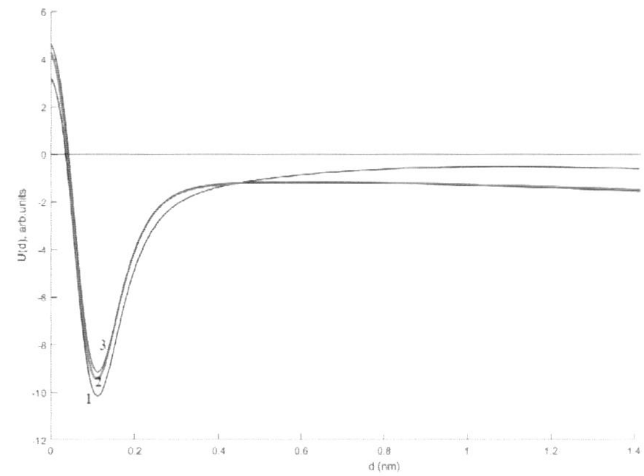

Figure 5. Interaction potential with different curves correspond to different values of the dielectric constants and surface roughness (curve 1: substrate III; curve 2: substrate I; curve 3: substrate II)

4. Conclusion

In our work, we have discussed the effect of the interaction potential between nanoparticle and the metallic substrate using the method of image in dyadic Green's function. First, we discuss the case without surface roughness displaying through the field plots. According to the theory, the field depended on that the dielectric constant of each substrate was increased. As it has shown from contour graphs, the images particle under substrate with a high dielectric constant will more intensity than low dielectric constant. Because of high dielectric constant material is more the ability of the material to store electrical energy than the others. When we include with roughness consideration, one can see the decreasing of probability in substrate II and corresponds to the potential in Fig. 5 which are caused by surface morphology. The interaction is in addition to dielectric constant dependent on the substrate, moreover physical properties of the substrate is significant to investigate.

5. Reference

[1] Khudik B I and Lozovski V 1990 *Phys. Status Solidi B* **158** 511
[2] Keller O 1996 *Phys. Rep.* **268** 85
[3] Novotny L and Hecht B 2006 *Principles of nano-optics* (New York: Cambridge University Press) p 25
[4] Kysylychyn D, Piatnytsia V and Lozovski V 2013 *Phys. Rev. E* **88** 052403
[5] Albanese A, Tang P S and Chan W C W 2012 *Annu. Rev. Biomed. Eng.* **14** 1
[6] Keller O 2011 *Quantum Theory of Near-Field Electrodynamics* (New York: Springer) p 53
[7] Sipe J E, Young J F, Preston J S and van Driel H M 1982 *Phys. Rev. B* **27** 1141
[8] Girard C, Joachim C and Gauthier S 2000 *Rep. Prog. Phys.* **63** 893
[9] Keller O, Xiao M and Bozhevolnyi S 1993 *Surf. Sci.* **280** 217
[10] Ohtsu M 1998 *Near-Field Nano/Atom Optics and Technology* (Yokohama: Springer) p 273
[11] Boyd R W 2003 *Nonlinear Optics* vol 2 (San Diego: Academic Press) p 4
[12] Keller O 1986 *Phys. Rev. B* **34** 3883

IOP Conf. Series: Materials Science and Engineering **526** (2019) 012017 doi:10.1088/1757-899X/526/1/012017

Chirality Effects on an Electron Transport in Single-walled Carbon Nanotube Coupling to a Quantum Dot

S Boonchui[1,2,*] and J Charoenpakdee[1]

[1] Department of Physics, Kasetsart University, Bangkok, 10900, Thailand.
[2] Specialized center of Rubber and Polymer Materials in agriculture and industry (RPM), Faculty of Science, Kasetsart University, Bangkok, 10900, Thailand.

*Corresponding author: fscistb@email.com

Abstract. We investigated the effects of asymmetric velocities of Dirac particles, calculated by inducing curvature-induced σ-π mixing and Slater-Koster type projection for π-π and σ-π hopping integrals, on the transfer energy between the quantum dot (QD) and electron in a carbon nanotube. We also showed that probability for finding the electron on the left-side is more than the right-side of carbon nanotube. This result corresponds with the asymmetric velocities of Dirac particles, such a study may prove a useful for charge transport of QD-solar cell.

1. Introduction

Carbon Nanotube (CNT) photonics with fascinating properties [1] is an emerging field. It offers a challenge to both fundamental and applied sciences. Many properties of CNT are tied to its lattice structure, allowing for tuning of charge carrier dynamics through the curvature of the nanotube surface, physical scenarios in condensed matter setting: narrow-linewidth blackbody emission [2], curvature effect on polarization of light emitted from chiral carbon nanotubes [3], effect of curvature-induced superlattice structures on energy band structures of helically coiled carbon nanotubes [4], etc. The efficiency of absorption and emission of light depends on different chirality of the individualized nanotubes [5]. The photoluminescence (PL) intensity of a single-wall carbon nanotube (SWCNT) is calculated by multiplying the photon-absorption, relaxation and photon-emission matrix elements [6]. In 2012, W. Izumida *et al.* showed the velocities of Dirac particle are asymmetric due to a second order perturbation technique of curvature-induced σ - π mixing with Slater-Koster type projection for π - π and σ - π hopping integrals on the curved surface of nanotubes [7]. Concerning to the hexagonal lattice and graphene, the current-current correlation function method is used [8]. The correlation function can be decomposed into parallel and perpendicular contributions.

The present paper is organized as follows: we define the continuum model and derive the current local operator of this model (Sec. II), we further show the calculation of probability amplitude which decays the excitation state and has the electron at either right-side or left-side of the carbon nanotube. In Sec. III, we analyze the numerical results and discussion of the scattering of electron, depending on the asymmetric velocities of Dirac particles. A summary, along with some concluding remarks, is given in Sec. IV.

IOP Conf. Series: Materials Science and Engineering **526** (2019) 012017 doi:10.1088/1757-899X/526/1/012017

2. Time evolution of the right-going and left-going electrons

We investigate the scattering of a single electron in chirality single-walled carbon nanotube (C-SWCNT) interacting with a V-type three-level quantum dot (V-QD), coupled to 1D electron transport which is a metal nanowire and we consider an electron in near Dirac point region, respectively.
First of all, we study the scattering of electron interaction with V-QD as shown in Fig. 1. In our model, the composite system is considered as an electron transport-quantum dot interaction, and the Hamiltonian system is given by

$$\hat{H} = \hat{H}_{ele} + \hat{H}_{qd} + \hat{H}_{\text{int}} \,. \tag{1}$$

where $\hat{H}_{ele} = \sum_{\gamma=R,L} \int dz\, \hbar(-iv_F^{(\gamma)} c_\gamma^\dagger(z)\frac{\partial}{\partial z}c_\gamma(z) + H.c.)$ is both right-going and left-going Hamiltonian of carbon nanotube, $c_R^\dagger(z)$ and [$c_L^\dagger(z)$] are creation operator of electron a right- [left-] going at z , $C_R(z)$ and [$C_L(z)$] are annihilates a right-[left-] going at z, $v_F^{(\gamma)}$ is the group velocity of an electron transport in C-SWCNT for the right-going $v_F^{(R)}$ and left-going $v_F^{(L)}$ describing the creation operation. The Hamiltonian describes the Λ-type three-level structure:

$$\hat{H}_{qd} = \hbar(\omega_{10} - i\Gamma_1)|1\rangle\langle 1| + \hbar(\omega_{20} - i\Gamma_2)|2\rangle\langle 2| + \hbar\omega_{00}|0\rangle\langle 0| \,. \tag{2}$$

where $|0\rangle$ is ground state of quantum dot and $\Gamma_1[\Gamma_2]$ is constant of decay of electron in quantum dot at excited 1[2] state. The electron transport-quantum dot coupling Hamiltonian is given by:

$$\hat{H}_{\text{int}} = \sum_{\gamma=R,L}\sum_{\eta=1,2} V_\gamma^{(\eta)}\hbar\int dz\, \delta(z)(c_\gamma^\dagger(z)\hat{\sigma}_\eta^{(-)} + H.c.) \tag{3}$$

where $V_\gamma^{(\eta)} \alpha\, v_F^{(\gamma)} |\vec{d}_{\eta 0}|$ is coupling constants, which is proportional to the transition dipole moment $|\vec{d}_{\eta 0}|$ and $v_F^{(\gamma)}$ the group velocity of an electron transport, $V_\gamma^{(1)}$ is time independent but $V_\gamma^{(2)}$ is time dependent. Effects of the group velocity on photon emission of the dark ($|2\rangle$) and the bright ($|1\rangle$) exciton states have been investigated.

The studied system is modeled as shown in Fig. 1, the V-QD has been exited to the excitation state, $|1\rangle$ or $|2\rangle$. Then, the excitation state decay and has the transitions $|1\rangle \leftrightarrow |0_{qd}\rangle$ or $|2\rangle \rightarrow |0_{qd}\rangle$ with resonance frequencies (dipole transition) ω_{10} and ω_{20}, accompanying the transition dipole moment ($\vec{d}_{10}$) and ($\vec{d}_{20}$) respectively.

Energy is transferred to C-SWCNT, creating both right-going and left-going electrons in the carbon nanotube. We investigate the effects of asymmetric velocities of Dirac particles on the probability amplitude for finding at the right-side and left-side of the carbon nanotube, given by $Amp_L^{(\eta)} = \langle L|\exp(-i\hat{H}t/\hbar)|in\rangle$ | and $Amp_R^{(\eta)} = \langle R|\exp(-i\hat{H}t/\hbar)|in\rangle$, respectively.

where $|L\rangle = \theta(-x)e^{-\frac{i}{\hbar}\hat{H}_{ele}t}\hat{c}_L^\dagger(x)|0,0_{qd}\rangle$ and $|R\rangle = \theta(x)e^{-\frac{i}{\hbar}\hat{H}_{ele}t}\hat{c}_R^\dagger(x)|0,0_{qd}\rangle$ are a quantum state for finding the electron at the right-side and left-side of the carbon nanotube, respectively, $|0,0_{qd}\rangle$ is the V-QD at ground state and the electron in the carbon nanotube is in vacuum state. In Fig. 1, the initial state of system can be assumed that $|in\rangle = |0,\eta\rangle$ is the V-QD at excitation state, $|\eta\rangle$; $\eta = 1,2$ and the electron in carbon nanotube is in vacuum state.

For calculation of probability amplitude $\langle L|\exp(-i\hat{H}t/\hbar)|in\rangle$ and $\langle R|\exp(-i\hat{H}t/\hbar)|in\rangle$, we keep only the terms of first-order in $V_\gamma^{(\eta)}$ and we then obtain can be given as:

$$Amp_L^{(\eta)}(x,t) = \langle L|e^{-\frac{i}{\hbar}\hat{H}t}|in\rangle \approx \theta(-x)\langle 0_{qd}|\hat{c}_L(x)\sum_{n=0}^{\infty}\frac{(-i)^n\hbar^n}{n!\hbar!}\int_0^t dt'\cdots\int_0^{t^{(n-1)}} dt^{[n]}\left(\omega_{\eta 0}-\frac{i\Gamma_{\eta 0}}{2}\right)^n\left[\frac{V^L}{\sqrt{2}}\hat{c}_L^\dagger(0,t'')+\frac{V^R}{\sqrt{2}}\hat{c}_R^\dagger(0,t'')\right]|\eta\rangle \quad (4)$$

and

$$Amp_R^{(\eta)}(x,t) = \langle R|e^{-\frac{i}{\hbar}\hat{H}t}|in\rangle \approx \theta(x)\langle 0_{qd}|\hat{c}_R(x)\sum_{n=0}^{\infty}\frac{(-i)^n\hbar^n}{n!\hbar!}\int_0^t dt'\cdots\int_0^{t^{(n-1)}} dt^{[n]}\left(\omega_{\eta 0}-\frac{i\Gamma_{\eta 0}}{2}\right)^n\left[\frac{V^L}{\sqrt{2}}\hat{c}_L^\dagger(0,t'')+\frac{V^R}{\sqrt{2}}\hat{c}_R^\dagger(0,t'')\right]|\eta\rangle \quad (5)$$

We can calculate the probabilities for finding the electron at the right and left of the carbon nanotube by using the creation operator of the electron a right- and left-going as defined [14]:

$$\hat{c}_R^\dagger(0,t) = \int_0^\infty \frac{dk}{2\pi}e^{-i\omega_k t}\hat{b}_k^\dagger \text{ and } \hat{c}_L^\dagger(0,t) = \int_{-\infty}^0 \frac{dk}{2\pi}e^{-i\omega_k t}\hat{b}_k^\dagger . \quad (6)$$

where $\hat{b}_k^\dagger$ is a creation operator of electron in momentum space to be the Fourier transforms of real-space operators.

3. Calculation and Discussion

In order to compute the correlation function with the diameter of SWCNTs and the chiral angle as the zigzag SWNT, (9,0) and the chiral SWNT (7,1) and the armchair SWNT, (4,4). The armchair (4,4) is gapless but the energy gap appears for the zigzag (9,0) and the chiral (7,1), the group velocity has the asymptotic velocity of the conduction and valence bands having a linear dispersion. Here the velocities are given by $v_F^L = -\partial E_L/\hbar\partial k, v_F^R = \partial E_R/\hbar\partial k$ where $E_{L(R)}$ is the energy band of the left-going (right-going) particle, which has the negative (positive) slope. The ratio of velocities to be $|v_F^R/v_F^L|$ are about 0.8, 0.94, and 0.75 for the armchair (4,4), the chiral (7,1), and the zigzag (9,0), respectively. We numerically calculate Eq. (5) by following parameters in ref. [9] as shown in Fig. 2 and Fig. 3

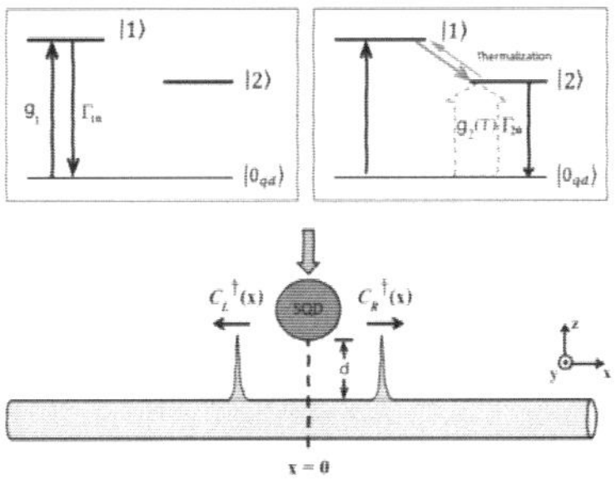

Fig. 1 Schematic diagrams of SWCNT directly coupled to the exciton state $|1\rangle$ (upper left) and virtually coupled to the exciton state $|2\rangle$, through thermalization, (upper right). (Bottom) Propagating electrons in the nanowire are created by decaying the exciton state.

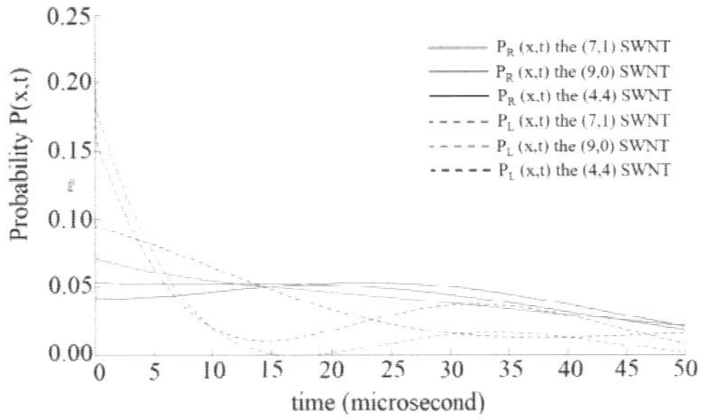

Fig. 2 Show that numerical results of probability for finding the electron at the right and left of the carbon nanotube are the function of time for the position x fixed. A probability for finding the electron at the left (Dot line) is more than the right (Solid line).

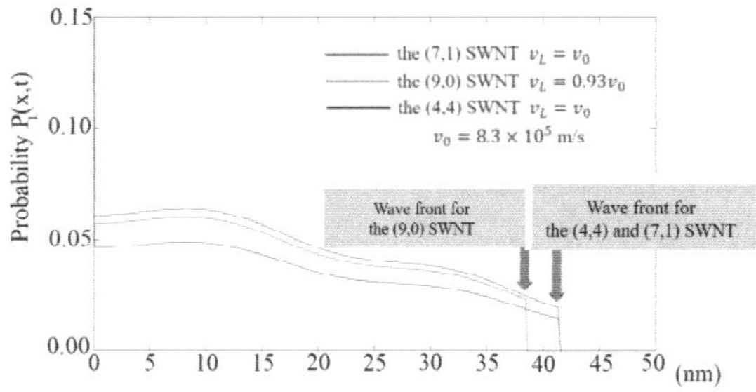

Fig. 3 Show that numerical results of probability for finding the electron at the left of the carbon nanotube is a function of position for time t fixed. A probability for finding the electron at the left of SWCNT correspond with the velocities, given by defining $v_F^L = -\partial E_L / \hbar \partial k$.

4. Conclusions

We have provided a quantum mechanical theoretical study of the stimulated electron which has right- or left-going in SWCNT. Numerical results of probability for finding the electron at the right-side and left-side of the carbon nanotube show that a probability for finding the electron at the left-side of the carbon nanotube is a function of position for time *t* fixed. Probability for finding the electron at the left-side SWCNT corresponds to the velocities, given by defining $v_F^L = -\partial E_L / \hbar \partial k$. This study provides insights into this fundamental process of nature, including the role of lattice structure on the probability of creation of the right- and left-going electrons.

5. References

[1] R. Miura, S. Imamura, R. Ohta, A. Ishii, X. Liu, T. Shimada, S. Iwamoto, Y. Arakawa and Y. K. Kato, Ultralow mode-volume photonic crystal nanobeam cavities for high-efficiency coupling to individual carbon nanotube emitters, Nat. Commun. **5**, 5580 (2014).

[2] T. Mori, Y. Yamauchi, S. Honda, and H. Maki, An Electrically driven, ultrahigh-Speed, on-chip light emitter based on carbon nanotubes, Nano Lett. **14**, 3277 (2014).

[3] Thiti Thitapura, Watchara Liewrian, Tula Juarosaga, and S. Boonchui, Curvature effect on polarization of light emitted from chiral carbon nanotubes, Optics Express **25**, 25588 (2017).

[4] Thiti Thitapura, Watchara Liewrian, Tula Juarosaga, and S. Boonchui, Effect of curvature-induced superlattice structures on energy band structures of helically coiled carbon nanotubes, Plasmonics **12**, 1439(2017).

[5] C. Fantini, J. Cassimiro, V.S.T. Peressinotto, F. Plentz, A.G. Souza Filho, C.A. Furtado and A.P. Santos,"Investigation of the light emission efficiency of single-wall carbon nanotubes wrapped with different surfactants," Chem. Phys.Lett. **473**, 96 (2009).

[6] W. Izumida, A. VikstroÜm, and R. Saito, "Asymmetric velocities of Dirac particles and Vernier spectrum in metallic single-wall carbon nanotubes," Phy. Rev. B **85**, 165430 (2012).

[7] T. Stauber and G. Gómez-Santos, Dynamical current-current correlation of the hexagonal lattice and grapheme, Phys. Rev. B **82**, 155412 (2010).

[8] P. Srisangyingcharoen, R. Klinkla and S. Boonchui, Dynamics of propagating surface plasmon induced photon emission from quantum dots: quantum history approach, J. Phys. B: At. Mol. Opt. Phys. **48**, 215501 (2015).

IOP Conf. Series: Materials Science and Engineering **526** (2019) 012018 doi:10.1088/1757-899X/526/1/012018

Effects of Sn Incorporation in ZnO Thin Films on Properties of Perovskite Solar Cells

P Malison[1,2], C Bhoomanee[1,2], S Choopun[1,2], D Wongratanaphisan[1,2], T Sagawa[3], and P Ruankham [1,2,*]

[1] Department of Physics and Materials Science, Faculty of Science, Chiang Mai University, Chiang Mai, 50200, Thailand
[2] Thailand Center of Excellence in Physics (ThEP center), CHE, Ratchathewi, Bangkok, 10400, Thailand
[3] Graduate School of Energy Science, Kyoto University, Yoshida-Honmachi, Sakyo-Ku, Kyoto 606-8501, Japan

* Corresponding author: pipat.r@cmu.ac.th

Abstract. Properties of electron transporting layer (ETL) play an important role on photovoltaic performances of perovskite solar cells. In this work, effects of Sn incorporation on properties of ZnO-based perovskite solar cells were investigated. Sn-doped ZnO (TZO) thin film as ETL was prepared via a sol–gel method. With 5% atom doping, TZO film coated on an indium doped tin oxide (ITO) substrate provided comparable light transmittance with that of an undoped ZnO/ITO substrate. It was also found that the optical band gap of TZO film (3.30 eV) is slightly wider than that of the ZnO one (3.28 eV). These results suggest that Sn atoms probably incorporated into the ZnO crystal during the sol-gel method. The grains size of perovskite layer coated on TZO or ZnO films also showed variation. The perovskite crystal on the TZO thin film (average 300 nm) was larger than that of the one on ZnO thin film (average 277 nm). The preliminary results indicate that the perovskite solar cell based on TZO film provided higher power conversion efficiency (PCE) of 4.42 % than the ZnO-based device (3.16%). Short-circuit current density (J_{sc}), open-circuit voltage (V_{oc}) and fill factor (FF) of TZO-based device were also higher than the ZnO-based device. This may be because TZO film may provide lower resistivity and better ETL/perovskite interface contact, confirmed by lower series resistance and higher shunt resistance of the TZO-based device. Finally, this work introduced a simple method to prepare TZO film at low temperature for photovoltaic application. It may help guide the development of flexible solar cells and other optoelectronic devices.

1.Introduction

Solar energy is a very important energy source because it is renewable, inexpensive, and clean [1]. Perovskite solar cell (PSC) is an emerging type of solar cells and it is very promising for future light-to-electricity conversion. Since the past few years, the PSCs have been developed rapidly because of their excellent semiconducting properties; tunable band gap [2], simple deposition process, low-cost[3], wide-range light absorption, long carrier diffusion length, and high carrier mobility [3].

PSCs can be constructed by sandwiching a perovskite layer with electron transporting layer (ETL) and hole transporting later (HTL). Zinc oxide (ZnO) is one of the materials using as ETL [4] because it provides good properties of n-type semiconductor such as wide-direct band-gap energy (3.37 eV), high exciton binding energy, non-toxicity, high transparency, and applicability for flexible devices [4-5].

In order to prepare flexible substrates, the ZnO ETL is required to be processed at low temperature due to low heat resistance of plastic substrate [5]. However, the electron mobility of ZnO film prepared at low temperature is relatively low in comparison to the film processed at higher temperature [4]. This film may result in poor device performance [4]. Therefore, ZnO should be modified by introducing element dopants to enhance its electrical properties [7-9].

In this work, Sn atoms were incorporated into the ZnO thin films via a sol-gel method. The molar ratio of Sn/(Zn + Sn) was fixed at 5% since this ratio provided the optimum improvement in film conductivity.[10] The optical properties of Sn-doped ZnO (TZO) thin film were also investigated. The improvement in photovoltaic performances was achieved with TZO film was used as ETL.

2. Experimental

2.1 Preparation of TZO thin films

An indium tin oxide (ITO) glass substrate was patterned using Zn powder and 1 M of HCl solution. Then, the substrate was sonicated in 1% w/v Alconox solution, deionized water (DI), acetone and isopropanol for 30 min each. The cleaned ITO glass substrate was treated with UV-ozone cleaner for 30 min. A TZO precursor solution was prepared by dissolving $Zn(CH_3COO)_2.2H_2O$ and $SnCl_2 \cdot 2H_2O$ in 2-methoxyethanol with a total metal concentration of 0.50 mol/L. The mixture was stirred overnight at room temperature. The molar ratio of Sn/(Zn + Sn) was fixed at 5% [9]. Compact TZO layers were deposited onto the cleaned ITO substrates by spin-coating the precursor solution on the cleaned ITO substrate at 3,000 rpm for 60s. The substrate was then annealed on a hot plate at 160 °C for 1 h.

2.2 Preparation of device

A PCBM layer was coated on the TZO layer by spin-coating a PCBM solution in chlorobenzene at 2000 rpm for 30 s. The films were annealed at 70 °C for 5 min. A perovskite layer was prepared via two-step sequential deposition [11]. A 1 M of PbI_2 solution in N,N-Dimethyl formamide (DMF) with 120 µl of 4-*tert*-butylpyridine (tBP) was deposited on the TZO substrate by spin-coating at 3,000 rpm for 30s and annealing on a hotplate at 100 °C for 60 min. After cooling to room temperature, a mixture of FAI ($HN{=}CHNH_3I$) and MAI (CH_3NH_3I) with molar ratio of 4:6 in isopropanol was dropped on the PbI_2 layer in a N_2-filled glove box at relative humidity of less than 20%. Then, it was spin-coated at 2,000 rpm for 30s. The as-deposited film was annealed on a hotplate at 75 °C for 20 min to form $MA_{0.6}FA_{0.4}PbI_3$. A dopant-free poly (3-hexylthiophene-2,5-diyl) or P3HT layer was formed by spin-coating a 20 mg/ml P3HT solution in chlorobenzene at 1,000 rpm for 30s. Finally, Au metal electrode was deposited by thermal evaporation on the P3HT layer.

2.3 Characterizations

UV–vis spectroscope (Lambda35 spectrophotometer) was used to investigate optical properties. The average transmittance was defined as an arithmetic mean of the transmittance values over 300-800 nm [12]. Field emission scanning electron microscope (FE-SEM) was carried out to observe and characterize perovskite morphology. The FE-SEM images were analyzed by randomly measuring the sizes of 130 grains using ImageJ software in order to determine grain size distribution. Photocurrent-voltage measurement under solar light AM 1.5 at 1,000 mW/cm^2 was carried out to characterize photovoltaic properties of perovskite solar cells. A photomask with area of 0.038 cm^2 was used to define irradiated active area.

3. Results and Discussion

To investigate the effects of Sn incorporation on optical properties, transmittance and absorbance of TZO films were measured. The transmission spectra of ZnO and TZO thin films on ITO substrates were shown in figure 1a while the absorption spectra were shown in figure 1b. It is found that the average transmittances are 78.47 % for ZnO film and 77.94 % for TZO one. The absorbance of the TZO film is significantly lower than that of the ZnO film. These results may be affected by the Sn incorporation.

In addition, the energy gap (E_g) of these films was calculated in order to roughly confirm the presence of Sn atoms in the ZnO film. The optical bandgap of the films was determined by plotting $(\alpha h\upsilon)^2$ as a function of the photon energy ($h\upsilon$), as shown in figure 1c. Note that α, h, and υ are absorption coefficient, Plank's constant and photon frequency, respectively.

The E_g values of these films were determined by the following equation;

$$(\alpha hv) = \beta(hv - E_g)^m$$

where m is constant which depends on types of optical transition. In this case, since ZnO is direct band gap material, m is 1/2. The extrapolation of this linear line intercepted the x-axis gave the E_g value. The E_g of ZnO film and TZO films were approximately 3.28 and 3.30 eV, respectively. The change in E_g values could be influenced by the Sn incorporation.

The perovskite layers were deposited on ZnO and TZO films by two step deposition technique. FE-SEM images of perovskite flms on those samples were showed in Figure 2. The average grain size of perovskite materials on TZO film is approximately 302 nm and that on ZnO film is approximately 277 nm.

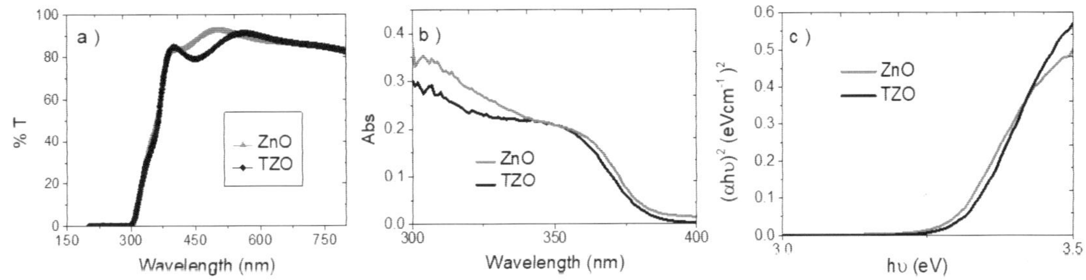

Figure 1. a) Transmittion spectra, b) absorption spectra, and c) Tauc plots of TZO and ZnO thin films.

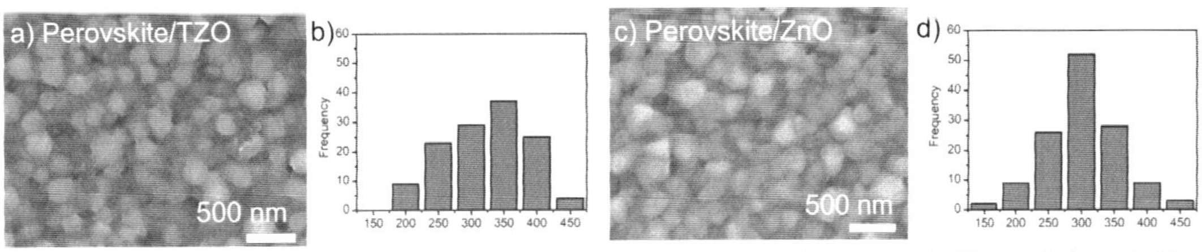

Figure 2. Top view FE-SEM images of perovskite layers prepated on a) TZO film and c) ZnO film. b) and d) showed grain size distribution of perovskite layer on TZO and ZnO respectively.

To further verify the Sn-doping effects on photovoltaic performances of perovskite solar cells, photocurrent-voltage (J-V) measurement was performed. The photovoltaic parameters were evaluated and summarized in Table 1. The perovskite solar cell based on TZO film provided an average power convertion efficiency (PCE) of 4.19 % (maximum PCE of 4.42%). These values are higher than that of the device based on ZnO film (average PCE of 3.11% abd maximum PCE of 3.16%). It was also found that short-circuit current density (J_{sc}), open-circuit voltage (V_{oc}) and fil factor (FF) of the TZO device are higher than those of the ZnO device. The reason behind this improvemet can be attibuted to higher shint resistance (R_{sh}) and lower series resistance (R_s) of the TZO device over the ZnO one. These results imply that the contact resistances between the active layer and electrodes was improved and charge transport may be facilitated due to the Sn incorporation.

Table 1. The average photovoltaic parameters of perovskite solar cells fabricated in this work. The maximum PCE is shown in the parentheses.

ETM	V_{oc} (V)	J_{sc} (mA cm^{-2})	FF	PCE (%)	R_{sh} (Ωcm^2)	R_s (Ωcm^2)
TZO	0.94	9.20	0.48	4.19(4.42)	78	16
ZnO	0.87	8.59	0.42	3.11(3.16)	340	28

4. Conclusions

Sn atoms could be incorporated into ZnO thin film, forming TZO film. The perovskite solar cell based on TZO film provide the enhanced photovoltaic performances. This improvement is due to better interfacial property and charge transport. Finally, this work introduced a simple method to prepare TZO film at low temperature for photovoltaic application. It may help guide the development of flexible solar cells and other optoelectronic devices.

5. References

[1] Angmo D and Krebs F C 2015 *Energy Technol.* **3** 774-83
[2] Huang H, Raith J, Kershaw S V, Kalytchuk S, Tomanec O, Jing L, Susha A S, Zboril R, and Rogach A L 2017 *Nat. Commun.* **8** 996
[3] Jia X, Wu N, Wei J, Zhang L, Luo Q, Bao Z, Li Y Q, Yang Y, Liu X, and Ma C Q 2016 *Org. Electron.* **38** 150-7
[4] Wei J, Yin Z, Chen S C, and Zheng Q 2017 *ACS Appl. Mater. Interfaces* **9** 6186-93
[5] Muslih E Y and Kim K H 2017 *IOP Conf. Ser.: Mater. Sci. Eng.* **214** 012001
[6] Fakharuddin A, Jose R, and Brown TM 2016 *Energy Environ. Sci.* **9** 3007-35
[7] An Q, Fassl P, Hofstetter YJ, Becker-Koch D, Bausch A, Hopkinson PE, and Vaynzof Y 2017 *Nano Energy* **39** 400-8
[8] Naz H, Ali RN, Liu Q, Yang S, and Xiang B 2018 *J. Colloid. Interface Sci.* **512** 548-54
[9] El Mir L 2017 *J. Lumin.* **186** 98-102
[10] Venkataraj S, Hishita S, Adachi Y, Sakaguchi I, Matsumoto K, Saito N, Haneda H, and Ohashi N 2009 *J. Electrochem. Soc.* **156** H424-H9
[11] Mahmud MA, Elumalai NK, Upama MB, Wang D, Wright M, Chan KH, Xu C, Haque F, and Uddin A 2016 *Electrochim. Acta* **222** 1510-1521.
[12] Hu D, Cao J, Li W, Zhang C, Wu T, Li Q, Chen Z, Wang Y, and Gaun J 2017 *Adv. Optical Mater.* 1700109

Acknowledgments

This research was financially supported by Chiang Mai University and the Development and Promotion of Science and Technology Talents Project (DPST) (Research Fund for DPST Graduate with First Placement No. 25/2557). The authors would like to acknowledge the Science Achievement Scholarship of Thailand and Graduate School Chiang Mai University. P Malison would like to acknowledge Research Professional Development Project Under the Science Achievement Scholarship of Thailand.

IOP Conf. Series: Materials Science and Engineering **526** (2019) 012019 doi:10.1088/1757-899X/526/1/012019

Extraction of nanocellulose from pineapple leaves by acid-hydrolysis and pressurized acid hydrolysis for reinforcement in natural rubber composites

W Chawalitsakunchai[1], P Dittanet[1], S Loykulnunt[2], S Tanpichai[3], and P Prapainainar[1,*]

[1]National Center of Excellence for Petroleum, Petrochemicals and Advance Material, Department of Chemical Engineering, Faculty of Engineering, Kasetsart University, Bangkok 10900, Thailand
[2]Natural Rubber Focus Unit, National Metal and Materials Technology Center, NSTDA, 111 Paholyothin Rd., Klong 1, Klong Luang, Pathumthani 12120 Thailand
[3] Learning Institute and Cellulose and Bio-based Nanomaterials Research Group,, King Mongkut's University of Technology Thonburi, Bangkok, 10140, Thailand

* Corresponding author: fengpwn@ku.ac.th

Abstract. This work aimed to study the extraction of nanocellulose (CNC) from pineapple leaves (PL) for using as a reinforcing filler in the natural rubber. PL mainly consists of cellulose so it is properly used for nanocellulosic extraction. The extractions were performed by acid-hydrolysis and pressurized acid-hydrolysis methods. The morphology and structure of the extracted nanocelluloses were compared and investigated using transmission electron microscope, Fourier transformed infrared, and X-ray diffraction analysis. It was found that pineapple leave has 27.64% of α-cellulose. The size of celluloses, which were extracted and hydrolyzed by acid- hydrolysis and pressurized acid hydrolysis, were reduced to 130 and 117 nm, respectively. And the shape was rod-like. The crystallinity index (CrI) of CNC was increased to 92.95 % and 91.24 % when it was treated by acid hydrolysis and pressurized acid hydrolysis, respectively.

1. Introduction

Pineapple is one of the most popular tropical fruits but only its fruit part is focused in food industry. The other parts especially leaves are considered as agricultural waste which is abundant and inexpensive. Its main chemical composition is cellulosic compound [1]. Cellulosic is well known and attractive material due to their various extraordinary properties such as high mechanical properties, biocompatibility and biological degradability [2, 3]. The size of micro cellulose (MCC) was reduced into nanocellulose (CNC), which retain unique properties of cellulose with the smaller size leading to their higher incorporated ability with polymer matrices [4]. In this work, CNC was extracted from pineapple leaves by used alkali treatment followed by bleaching treatment. The extracted cellulose was used to synthesize the nanocellulose using two different methods which are the acid hydrolysis and pressurized acid hydrolysis.

These two extracted nanocelluloses were investigated by transmission electron microscopy, Fourier transformed infrared, and X-ray diffraction analysis. It was further used as reinforcing filler in the natural rubber.

2. Experimental

2.1 Materials
Pineapple leaves was supplied from amphoe Bang Saphan, Prachuap Khiri Khan in Thailand. 97% purity sodium hydroxide (NaOH) was purchased from Ajax Finechem. 10% (w/v) sodium chlorite ($NaClO_2$) was purchased from local industry in Thailand. Acetic acid (CH_3COOH) was purchased from Quality Reagent chemical. 98% sulfuric acid (H_2SO_4) was purchased from Merck, Germany. And NR latex (60%DRC) was provided by Rubber Authority of Thailand.

2.2 Extraction of cellulose
First, raw pineapple leaves were washed in fresh water then chopped into approximately 2 cm × 5cm, dried in oven at 90 ºC, and grinded into 250 μm pineapple leave powder. After that, the cellulose was extracted from pineapple leave powder by alkali treatment using 50 g of pineapple leave powder to 1 L of 1 M (4 w/v %) NaOH at 80 °C for 2 hrs under stirring condition. Then, bleaching treatment was carried out three times using 2.5 %w/v $NaClO_2$ at 100°C for 1 hr using the ratio of material to liquid was 1:20 (g/ml). At the end the cellulose was neutralize by distilled water and dried in oven at 70°C overnight.

2.3 Acid hydrolysis
The extracted cellulose was hydrolyzed to obtain CNC through acid hydrolysis method. First, 1 g of extracted cellulose was added into 10 ml of 64 %w/w sulfuric acid to form a suspension. It was continued hydrolyzed at 45 °C for 1 hr. At the end of reaction, the suspension product was filled with 5 fold water for stop reaction and then centrifuged at 8,000 rpm for 10 min to separate the gel suspension and remove the excess acid. The gel was adjusted to neutral by filtering through dialysis membrane and immersed in water. Finally, neutral gel was sonicated by ultrasonic sonicator.

2.4 Pressurized acid hydrolysis
The dried pineapple leave powder was treated with 2 %w/v NaOH at 80 °C for 1 hr under autoclave. Then alkali cellulose was bleached with $NaClO_2$ at 90 °C for 2 hrs. The bleaching was repeated a few times. Finally, the bleached cellulose was hydrolyzed with 10 %w/w sulfuric acid at 45 °C for 4 hrs.

2.5 Characterization
Determination of compositions such as holocellulose, α-cellulose, hemicellulose, lignin, extractives and ash by TAPPI standard methods and acid chlorite method of Browing were carried out [5, 6]. The Fourier-transformed infrared spectroscopy (FT-IR) was used to investigate functional group in samples. The XRD pattern of samples was recorded to calculate crystallinity index (CrI) by using $I_c = (I_{crys+am}-I_{am})\times 100/\ I_{crys+am}$ where I_C is the crystallinity index (CrI), $I_{crys+am}$ is parts of crystalline and amorphous at $2\theta = 22^o$, and I_{am} is the intensity peak of only amorphous region at 2θ around 18°. Transmission electron microscopy (TEM) was used to examine the shape and the size of extracted cellulose and nanocellulose.

3. Result and discussion

3.1 Chemical composition

The chemical composition of pineapple leave powder is shown in Table1. It consisted of 44.60% holocellulose that was classified to 27.64% α-cellulose, 16.96% hemicelluloses, 8.03% of lignin, 39.42% of extractives, and 5.80% ash.

Table1. Chemical compositions in pineapple leaves.

Chemical compositions	Percentage (%)	Standard
Holocellulose	44.60	Acid chlorite Method of Browing
α-cellulose	27.64	TAPPI T203
Hemicellulose	16.96	Holocellulose - α-cellulose
Lignin	8.03	TAPPI T222
Extractives	39.42	TAPPI T204, T264 and T264
Ash	5.80	TAPPI T211

3.2 Fourier-transformed infrared spectroscopy (FT-IR)

FT-IR spectroscopy was used to investigate the functional groups in samples including bleached and hydrolyzed cellulose to confirm the successful of treatment in each step. FTIR images are shown in Figure 1. It was found that cellulose existed in all samples. The bonds referred to peaks of celluloses were at 3328, 2911, and 1634 cm^{-1} which assigned to –OH stretching vibrations of hydroxyl group and aliphatic C–H stretching. Peak around 1620 cm^{-1} assigned to carboxylate groups that absorbed some water [7, 8]. The result could confirm that cellulose was remained in the sample after treatment in each step.

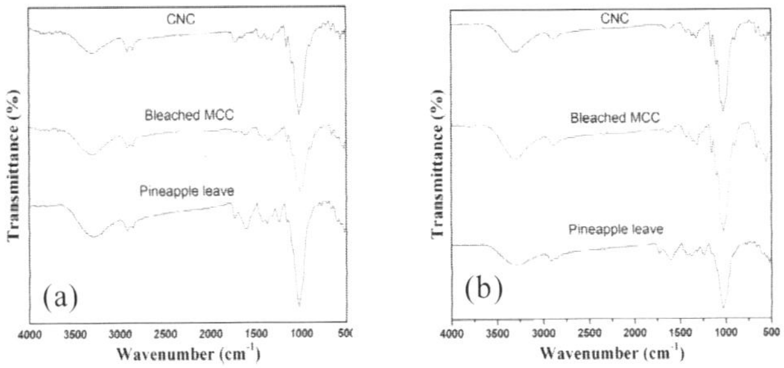

Figure 1. FTIR spectra image of pineapple leave, bleached MCC and CNC from acid hydrolysis (a) and pressurized acid hydrolysis (b).

3.3 X-ray diffraction (XRD)

XRD spectra exhibited the diffraction patterns of pineapple leave, bleached MCC, and hydrolyzed CNC (figure 2). At 2θ of around 18°, it referred to amorphous parts. At 2θ of around 22°, it assigned to amorphous and crystalline regions. The crystallinity index (CrI) of samples was calculated by Eq (1). The CrI of hydrolyzed CNC was increased to 92.95% and 91.24 % for acid hydrolysis and pressurized acid hydrolysis, respectively.

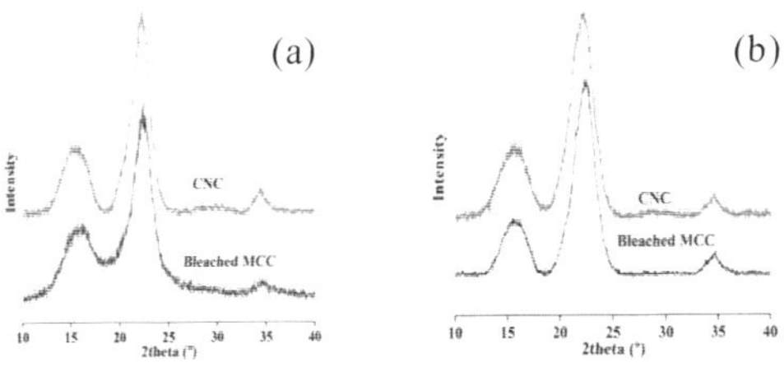

Figure 2. Image XRD patterns of MCC and CNC treated by acid hydrolysis (a) and pressurized acid hydrolysis (b).

3.4 Transmission electron microscopy (TEM)

The characteristic and shape of CNC hydrolyzed by the acid hydrolysis is shown in figure 3(a-c). The shape of CNC was found to be rod-like with the average length × width of 130.02 ± 48.55 nm × 5.14 ± 2.03 nm. In addition, CNC from pressurized acid hydrolysis is revealed in figure 3(d-e). The shape of CNC was similar to that hydrolyzed CNC. And average dimension was 3.52 ± 0.8 nm in width and 117.14 ± 24.75 nm in length.

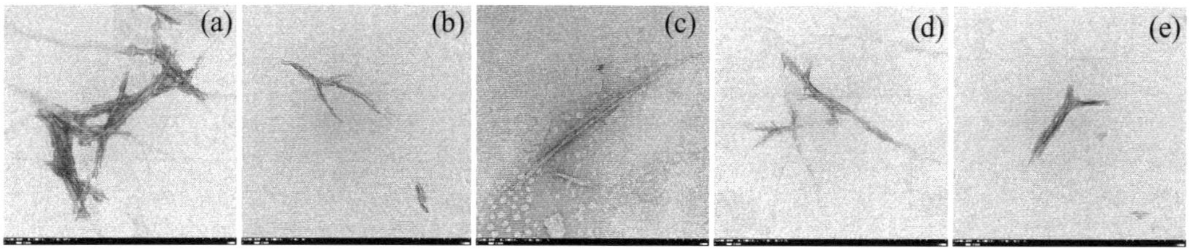

Figure 3. TEM images for CNC treated by acid hydrolysis (a - c) and pressurized acid hydrolysis (d - e).

4. Conclusion

This work studied cellulose extraction processes by alkali treatment and bleaching. CNC was hydrolyzed by acid hydrolysis and pressurized acid hydrolysis. The size of CNC was reduced into nano scale. It was found that the crystallinity index was increased. Therefore, CNC was suitable to be reinforcing agent in composites such as natural rubber. It will be used as filler in natural rubber composites for further study.

5. References

[1] Dos Santos R M, Flauzino Neto W P, Silvério H A, Martins D F, Dantas N O and Pasquini D 2013 *Industrial Crop and Products* **50** 707–714
[2] Qi H, Cai J, Zhang L and Kuga S 2009 *Biomacromolecules* **10** 1597–1602
[3] Naduparambath S, Jinitha T V, Shaniba V, Sreejith M P, Balan A K and Purushothaman E. 2018 *Carbohydr. Polym.* **180** 13–20
[4] Xu X, Liu F, Jiang L, Zhu J Y, Haagenson D and Wiesenborn D P 2013 *ACS Appl. Mater. Interfaces* **5** 2999–3009
[5] TAPPI Test Method, 2002–2003. The Technical Association of the Pulp and Paper Industry. TAPPI Press, Atlanta, Georgia
[6] Browning B L. 1967 Method of wood Chemistry Interscience Publisher
[7] Shaheen Th I and Emam H E 2018 *Carbohydr. Polym.* **107** 1599–1606
[8] Jiang F and Hsieh Y L 2013 *Carbohydr. Polym.* **95** 32–40

Acknowledgement

The authors are grateful that this work was funded by National Center of Excellence for Petroleum, Petrochemicals and Advance Material, Department of Chemical Engineering, Faculty of Engineering, Kasetsart University and Natioanl Nanotechnology Center (NANOTEC), NSTDA, Ministry of Science and Technology Center, Thailand, through its programme of Research Network NANOTEC (RNN).

Reduction of Endotoxin from Human Root Canals by Calcium Hydroxide Nanoparticles

P Lungkapinth[1], P Louwakul[1]

[1]Endodontic Section, Department of Restorative Dentistry and Periodontology, Faculty of Dentistry, Chiang Mai University, Chiang Mai 50200, Thailand.

* Corresponding author: pitchanun_l@cmu.ac.th

Abstract. This study was aimed to evaluate the reduction of endotoxin using the passive ultrasonic irrigation (PUI) of calcium hydroxide nanoparticles (CHNPs) compared with the PUI of calcium hydroxide (CH) and the CH medication. Forty-seven single-root human premolars were used. The root canals were enlarged and sterilized. *Escherichia coli* endotoxin was inoculated into the root canals. Thirty-six root canals were assigned into three experimental groups: CHNPs irrigation, CH irrigation, and CH medication. For the irrigation groups, the PUI was activated for three min. For the CH medication, the CH was filled into each canal and incubated at 37°C for seven days. The reduction of endotoxin was evaluated by quantitative chromogenic limulus amoebocyte lysate assay. Data was analysed by Kruskal-Wallis and pairwise comparison tests, with a level of significance set at $P < 0.05$. The CHNPs irrigation was more effective in against endotoxin than the CH irrigation ($P < 0.05$). No significant difference was found between the CHNPs irrigation and the CH medication ($P > 0.05$). The irrigation of CHNPs could eliminate most of the endotoxin from the root canals and approximately from 300 μm of root dentin. CHNPs might be potential as an adjunctive irrigation for endotoxin removal.

1. Introduction

Gram-negative bacteria are predominantly demonstrated through the infected pulp and apical periodontitis. They produce by-products and virulent factors that invade into periapical tissues such as lipopolysaccharides (LPS), or endotoxin, within their outer cell wall. LPS can induce inflammatory reaction and periapical bone resorption[1]. Thus, the main goal of root canal treatment in necrotic pulp with apical periodontitis is based on reduction or elimination of the microorganisms as well as inactivation of the endotoxin from the infected root canal, which lead to the healing of periapical tissues[2]. The chemo-mechanical procedures are effectiveness in reducing bacteria but there are not able to eliminate endotoxin from all root canals due to the irregularities of the root canal system[3]. Therefore, the intracanal medication or adjunctive approach has been recommended as an additional step for the root canal debridement. Passive ultrasonic irrigation (PUI), an adjunctive approach, which ultrasonic energy is transmitted through the root canal irrigant by the free movement of an activated file[4]. Previous studies have indicated that the PUI can reducing endotoxin in necrotic root canals[5, 6].

Calcium hydroxide (CH) is the most commonly used for intracanal medication, its use being for antimicrobial action, detoxifying LPS, and dissolving organic tissue debris. The effectiveness of CH can be limited by the buffering capacity of dentin, which affects the penetration of hydroxyl ions into dentinal tubules[7]. Nano-sized materials, with dimensions less than 100 nm diameter, have become interesting in dental applications. Higher charge density and surface areas of the nanoparticles enable greatly antibacterial efficacy[8]. Recently, Calcium hydroxide nanoparticles (CHNPs) were studied in endodontic researches such as its antimicrobial properties[8], penetration depth[9], and cytotoxicity[10]. However, the effect on the reduction of endotoxin has never been evaluated. Therefore, the aim of this study was to evaluate the reduction of endotoxin using the PUI of CHNPs compared with the PUI of CH and the CH medication.

2. Materials and Methods

The methods of this study were previously approved by the Human Experimentation Committee, Chiang Mai University, Thailand (protocol no. 014/2561). Forty-seven non-carious, single-rooted, permanent lower premolars which extracted for orthodontic reasons were used. The teeth were transversely cut at the cementoenamel junction with diamond discs. The roots were cut into fragments 10 mm long. The root canals were prepared by using peeso-drills (VDW GmbH, Munich, Germany) no #1 #2 and #3, respectively, and irrigated with 2.5% sodium hypochlorite after the use of each instrument. The roots were submerged and ultrasonicated in 17% EDTA for one min, 2.5% sodium hypochlorite for three min, and pyrogen-free saline for one min, respectively. The specimens were sealed at the apex and coated outer surfaces twice with nail varnish, except at the cervical opening.

All specimens were randomly divided into three experimental groups (n=12) and two control groups (n=9). The specimens were fixed with the chemical-cured acrylic resins in 24-cell culture plates (SPL, Pocheon-si, Gyeonggi-do, Korea). All specimens and equipment were previously sterilized by 60 Co gamma rays (Thailand Institute of Nuclear Technology, Nakornnayok, Thailand) at 20 KGy for six hours in order to eliminate the pre-existing endotoxin. Twenty microliters of *Escherichia coli* 055:B5 endotoxin (Sigma Aldrich, St Louis, MO, USA), at 400 EU/mL concentration, was subjected into each root canal, except in the negative control. The root canals were closed and vortexed for 15 min. All plates were incubated at 37±1°C in a 100% humidified atmosphere for 24 hours, then, the fillings were removed. Two teeth were selected for baseline samples.

2.1. Endotoxin removal procedures

2.1.1. PUI of CH and CHNPs. First, the root canals were irrigated with one mL of freshly prepared CH (Unilab; Ajax Finechem Pty. Ltd., Taren Point, NSW, Australia), 25 g/L dispersed in pyrogen-free water (A.N.B. Laboratories Co. Ltd., Kannayao Bangkok, Thailand), delivered by using disposable syringe with 22-gauge needle which was inserted to two mm short of root length. Second, the solution was activated by ultrasonic file #20/21 (IrriSafe tip; Satelec, Acteon Group, Merignec, France), driven by the ultrasonic device (Newtron®P5, Satelec) at a power setting of 4 for 30 s, with the probe tip placed at two mm short of root length. Then, each canal was refilled with another one mL of CH for 30 s and repeated the irrigation cycle twice to complete the irrigation in three min. The root canal was finally irrigated with five mL of pyrogen-free water for 60 s. To neutralize the activity of CH, the canal was rinse with five mL of 0.5% citric acid for 60 s and followed by five mL of pyrogen-free water for 60 s. Each root canal was dried, and the endotoxin sampling was collected. For the CHNPs group, the protocols were performed as same as the CH group which described above. The premixed form of CHNPs (CaLosiL® E25, IBZ-Salzchemie GmbH & Co.kg, Halsbrücke, Germany), 25 g/L CH dispersed in ethanol, was used instead of the CH and delivered by using disposable syringe with 27-gauge needle.

2.1.2. CH medication. CH (UltraCal XS; UltraDent Products, Inc., South Jordan, UT, USA), 35% in water, was delivered into each root canal by using a Navi-Tip (UltraDent Products, Inc.). All root canals were sealed with glass ionomer cement (GC Fuji IX GP®; GC America Inc., Alsip, IL, USA) and incubated at 37±1 °C in a 100% humidified atmosphere for seven days. When completed, the filling was removed and neutralized with five mL of 0.5% citric acid and followed by five mL of pyrogen-free water for 60 s. The endotoxin sampling was collected.

2.1.3. Positive control (n=6) and negative control (n=3). The specimens were divided into two subgroups according to the solvents of the irrigation used. In the pyrogen-free water group, the root canals were irrigated with 24 mL of pyrogen-free water for seven min, delivered by disposable syringe and 27-gauge needle at the same length. Another subgroup was irrigated with four ml of 75% v/v ethanol for three min followed by 20 mL of pyrogen-free water. In the negative control, the root canals were irrigated by 24 mL of pyrogen-free water for seven min.

Each canal was filled with pyrogen-free water, a pyrogen-free paper point was inserted into the full length of each canal for 60 s. The sampling was kept in the glass tube containing 1 mL of pyrogen-free water. Afterward, the root dentin was drilled by no #4 #5 and #6 peeso-drills (VDW GmbH) to assess the endotoxin from the different depths; 100, 200, and 300 μm. The endotoxin was extracted from the root dentin according to the method has been previously described[11]. All tubes were centrifuged for 20 min at 1000 rpm. The clear solution was collected and kept at temperature -20°C for the endotoxin analysis.

The quantitative chromogenic limulus amoebocyte lysate (LAL) assay was performed by using Pierce LAL chromogenic endotoxin quantitative kit (Thermo scientific Inc., Waltham, MA, USA) accoeding to the manufacturer's instructions. The endotoxin standard curve was created for calculation of the concentration of endotoxin. The coefficient of determination of the standard curve was 0.99. The test was performed in a 96-well microplate in a heating block at 37 °C throughout the assay. The plates were measured by using a spectrophotometer at an absorbance of 405 nm (Sunrise®; Tecan, Männedorf, Switzerland). The mean absorbance value of the blank was subtracted from the mean absorbance of the samples to calculate the mean absorbance. The endotoxin concentration (endotoxin unit per milliliters; EU/ml) was determined from the standard curve because the absorbance value was directly proportional to the concentration of endotoxin. All data was analyzed by Kruskal-Wallis and pairwise comparison tests. Significant level was considered at 5% ($P < 0.05$).

3. Results

Baseline samples were confirmed that the amount of endotoxin was distributed into the root canals. The mean values were 35.92 EU/mL, and the mean values of the root dentin layers at depth of 100, 200, and 300μm were 15.93, 10.51, 12.06 EU/mL, respectively. The reduction of endotoxin was extremely reduced after the PUI or the CH medication. Table 1 provided the values of endotoxin from the experimental groups. In the root canals, The PUI of CHNPs was more effective than the PUI of CH ($P < 0.05$). No significant difference between the PUI of CHNPs and CH medication ($P > 0.05$). In the root dentin, there was no statistically significant difference among the experimental groups at every depths of the root dentin ($P > 0.05$). The endotoxin was found in all levels of the samples in the positive control group, on the other hand, the endotoxin was not detected in the negative control group (data not shown).

Table 1. The median and range values of endotoxin concentration in the root canal samples collected from the experimental groups (EU/mL)

	PUI of CHNPs	PUI of CH	CH medication
		Median (range)	
Root canal	0.29 (0.02-0.75) ^{b}	0.70 (0.28-0.97) ^{a}	0.29 (0.02-0.75) ^{b}
Root dentin at 100 μm	0.05 (0.02-0.53) ^{A}	0.06 (0.04-0.46) ^{A}	0.06 (0.03-0.18) ^{A}
Root dentin at 200 μm	0.07 (0.04-0.39) ^{A}	0.08 (0.05-1.14) ^{A}	0.07 (0.03-0.24) ^{A}
Root dentin at 300 μm	0.08 (0.05-0.80) ^{A}	0.09 (0.05-0.47) ^{A}	0.06 (0.04-0.23) ^{A}

*Different letters indicate statistical significance for endotoxin between the different experimental groups (P < 0.05).
EU/mL = endotoxin unit per mililiters
PUI = passive ultrasonic irrigation, CHNPs = Calcium hydroxide nanoparticles, CH = Calcium hydroxide

4. Discussion

Endotoxin is released during death and lysis of gram-negative bacteria. The concentration of endotoxin may increase after the chemo-mechanical instrumentation[12] and remain in dentin smear layer[13]. It is difficult to eliminate the endotoxin from the root canals and dentin walls because of its irreversibly adherence to the mineralized tissues[14]. Gram-negative bacteria can penetrate into dentinal tubules at about 275 μm while

LPS represented approximately 4 times greater than the bacteria invasion according to its lower molecule weight[11]. The use of 400 EU/mL endotoxin in this study was based on the concentration found in the infected pulp with symptomatic apical periodontitis[15]. The chromogenic LAL test was used to measure endotoxin concentration, is based on the reaction between the clotting enzyme and endotoxin.

The CH medication for seven days can reduce the endotoxin only 4% from chemo-mechanical procedures[12], but, the CH medication in this study greatly reduced the endotoxin because the well-enlargement of the root canal may help the hydroxyl ions to diffuse into deep portions of the dentinal tubules.

The CHNPs solution is an alternative solution that is effective in against the endotoxin. The CHNPs solution used in this study contained nanoparticles of lime hydrate suspended in ethanol. The physical properties of CHNPs, such as the high solubility and stability in ethanol, the lower size of nanoparticles less than 100 nm may be attended to its neutralizing effect of endotoxin. The CHNPs may create more concentration of the hydroxyl ions that can break the fatty acid bonds of Lipid A[2]. The CHNPs were sufficiently penetrated to deep portions of dentinal tubules and increased charge density when reacting with remaining water in the dentin[8]. The nanoparticles solution may be able to diffuse greater than 300 µm depth of dentin. The PUI may also facilitate the penetration of the CHNPs solution because of the acoustic energy of the ultrasonic activation[4]. The well-replenished of solution during PUI may help to removed endotoxin from the root canals. Therefore, the use of CHNPs with PUI for three minutes may be beneficial in one-visit root canal treatment of the infected canals. Further studies are needed to evaluate the effectiveness of irrigation in eliminating endotoxin which adheres to the root dentin.

5. Conclusion
Within the limitations of this *in vitro* study, the PUI of CHNPs was almost eliminated endotoxin from the root canal and root dentin depth of 300 µm in straight root canals. CHNPs might be potential as an adjunctive irrigation for endotoxin removal.

6. References
[1] Pitts D L, Williams B L and Morton T H, Jr. 1982 *Journal of endodontics* **8** 10-8
[2] Xavier A C, Martinho F C, Chung A, Oliveira L D, Jorge A O, Valera M C and Carvalho C A 2013 *Journal of endodontics* **39** 959-64
[3] Martinho F C, Gomes A P, Fernandes A M, Ferreira N S, Endo M S, Freitas L F and Camoes I C 2014 *Journal of endodontics* **40** 625-9
[4] Van der Sluis L W, Versluis M, Wu M K and Wesselink P R 2007 *Int Endod J* **40** 415-26
[5] Nakamura V C, Pinheiro E T, Prado L C, Silveira A C, Carvalho A P L, Mayer M P A and Gavini G 2018 *Int Endod J* **51 Suppl 1** e12-e22
[6] Herrera D R, Martinho F C, de-Jesus-Soares A, Zaia A A, Ferraz C C R, Almeida J F A and Gomes B 2017 *Int Endod J* **50** 933-40
[7] Siqueira J F, Jr. and Lopes H P 1999 *Int Endod J* **32** 361-9
[8] Louwakul P, Saelo A and Khemaleelakul S 2017 *Clin Oral Investig* **21** 865-71
[9] Zand V, Mokhtari H, Hasani A and Jabbari G 2017 *Iran Endod J* **12** 366-70
[10] Dianat O, Azadnia S and Mozayeni M A 2015 *Iran Endod J* **10** 49-54
[11] Horiba N, Maekawa Y, Matsumoto T and Nakamura H 1990 *Journal of endodontics* **16** 331-4
[12] Vianna M E, Horz H P, Conrads G, Zaia A A, Souza-Filho F J and Gomes B P 2007 *Oral Microbiol Immunol* **22** 411-8
[13] Maekawa L E, Valera M C, Oliveira L D, Carvalho C A, Koga-Ito C Y and Jorge A O 2011 *J Appl Oral Sci* **19** 106-12
[14] Barthel C R, Levin L G, Reisner H M and Trope M 1997 *Int Endod J* **30** 155-9
[15] Martinho F C and Gomes B P 2008 *Journal of endodontics* **34** 268-72

Acknowledgements
The authors gratefully acknowledge the financial support provided by Chiang Mai University. The authors deny any conflicts of interest related to this study.

Potential Application of Biopolymer Chitosan and Cationic/Anionic Polymers in Textile Wastewater Treatment

W Watcharin[1] **and S Wiratthikowit**[1,*]

[1] Department of Agro-Industry, Faculty of Biotechnology, Assumption University, Hua Mak Campus, Bangkok 10240, Thailand.

* Corresponding author: waraleewtc@au.edu

Abstract. Biopolymer chitosan have been investigated for color removal in textile wastewater by using coagulation process of biopolymers and synthetic polymers. After the wastewater samples were treated with 0.1% chitosan for chitosan coagulation process, coagulation process of cationic or anionic polymer were performed by using chitosan and cationic polymer (CSC) or chitosan and anionic polymer (CSA) polymer concentrations of 0.1, 0.2 and 0.4 g/l at pH 6-8, respectively. The color removal efficiency of textile wastewater samples before and after treatment with chitosan followed by cationic/anionic polymers has been evaluated. The color removal by chitosan was dependent on pH and the maximum percent removal $31.4 \pm 3.3\%$ has been observed at pH 7. The optimum parameters for polymer-biopolymer coagulation process in this study was CSC at dosage of 0.2 g/l, pH 8 and CSA at dosage of 0.1 g/l, pH 7. Besides, its chemical analyses including Chemical Oxygen Demand (COD), dissolved solid, and suspended solid were analyzed in order to consider the potential application of chitosan in textile wastewater treatment. The results showed that the COD, dissolved solid and suspended solid of pre-treated textile wastewater were 290, 3998.7 and 280.7 mg/l, respectively, however, both dissolved solid and suspended solid were higher than the regulatory standard for wastewater. As the optimum condition, the COD, dissolved solid and suspended solid of textile wastewater after treatment with CSC (0.2 g/l at pH 8) were 260.7, 1939.8 and 103.6 mg/l, respectively, while the COD, dissolved solid and suspended solid of textile wastewater after treatment with CSA (0.1 g/l at pH 7) were 190.6, 3002.2and 106.8 mg/l, respectively. All of the results after treatments were within the regulatory standard. Therefore, the combination of chitosan and cationic/anionic polymers could be utilized in textile wastewater treatment.

1. Introduction

Nowadays, water pollution has gradually increased resulting from escalating pollution of fresh water resources either untreated or inadequately treated wastewater. The wastewater is a major environmental issue of the textile industries as textile wastewater pollutants are generally caustic soda, detergents, starch, wax, urea, ammonia, toxic dyes and heavy metal ions [1]. Because of threatened human health and ecological systems, the pollutant removal from textile wastewater is necessary. Therefore, various technologies in wastewater treatment have been studied such as adsorption, membrane filtration, chemical oxidation and coagulation [2]. The coagulation/adsorption is one of the most commonly used wastewater treatment techniques, which involved the physical adhesion of chemicals onto the surface of a solid.

Chitosan, the second most abundant biopolymer, are used for a wide range of applications i.e. agriculture, food processing, medicine, cosmetics, and biotechnology. For environmental applications, chitosan has been used as coagulant to treat a variety of wastewaters such as food processing industrial wastewaters, brewery wastewater, paper mill wastewater and olive oil wastewater [3]. As a biological cationic polymer, chitosan is being used to treat dairy wastewater at pH up to 5.25 by coagulation. Moreover, chitosan has a potential to remove pollutants containing metal ions and phenol derivatives. Chitosan with higher deacetylation degree exhibits greater efficiencies in removing pollutants either by adsorption or coagulation [4]. It was reported that the performance of chitosan in wastewater treatment vary with wastewater quality. For these reasons, this study was to investigate the potential application of chitosan for treatment of textile wastewater in the color removal, COD, DS and SS by coagulation-flocculation process.

2. Experimental

2.1. Materials

Textile wastewater was collected from Thai Towel Co. Ltd. (Nakornphatom, Thailand). The sample was stored at 4°C to minimize the changes in wastewater characteristics. Chitosan extracted from cuttlebones with a degree of deacetylation more than 80% and a moisture content of 9-10%, cationic polymer (FLOCTEX 3611) and anionic polymer (FLOCTEX 2413) were supplied by Thai Towel Co. Ltd. Glacial acetic acid, NaOH, HCl and other reagents were of analytical grade.

2.2. Coagulant Preparation

A conventional jar test was used in this study to coagulate wastewater using chitosan. 1 g/l chitosan dose was prepared by dissolving chitosan in 1% acetic acid. The experiment was carried out as a batch test, 600 ml of wastewater sample was treated with 15 ml chitosan solution under different operating pH (pH 6-8). The pH was controlled by adding either HCl or NaOH. After that, the wastewater suspension was agitated at various mixing time and speed, which consist of rapid mixing (100 rpm) for 1 min and slow mixing (30 rpm) for 30 min. After the agitation being stopped, the suspension was allowed to settle for 1 hour and the supernatant was collected for further analysis. After treated with chitosan, the parallel batch studies were performed using anionic or cationic polymer at dose of 0.1-0.4 g/l and pH 6-8 under the same agitation conditions to study the optimum condition for each parameter and their effect in coagulation and flocculation.

2.3. Analytical methods

The wastewater parameters such as pH, %transmittance at 550 nm, Chemical Oxygen Demand (COD), dissolved solid (DS) and suspended solid (SS) were analyzed according to the procedures described by the American Public Health Association (APHA). For color removal, the absorbance of the untreated and treated samples was measured at 550 nm using UV–VIS spectrophotometer. The optimum wavelength was obtained by scanning the untreated sample. The statistic analysis was performed by SigmaPlot using One Way ANOVA with Duncan's test.

3. Results and Discussion

The characteristics of the wastewater, such as an initial pH, %transmittance, COD, DS and SS were investigated as shown in Table 1 indicating the presence of high amounts of organic and inorganic matters in the textile wastewater. It can be noted that textile wastewater is generally alkaline and contains high pigment contents. Moreover, COD, DS and SS level in wastewater were used as the indicator on the sorption capacity to determine the optimum condition for the performance of chitosan in treatment.

Table 1. Characteristics of the textile wastewater sample.

Parameter	Value
pH	9
%transmittance	50.4 ± 0.5
COD, mg/l	290 ± 10
DS, mg/l	3,999 ± 101
SS, mg/l	281 ± 28

Many polyelectrolytes such as ionic polymers were widely used as chemical coagulants because they have high dye binding capacity, however, they are not easily biodegraded [5]. Chitosan are advantageous over chemical coagulants as it is non-toxic and biodegradable materials.

The percentage of color removal was investigated under different pH (6–8) using chitosan as a coagulant to treat textile wastewater shown in Figure 1. It can be seen that the color removal by chitosan was dependent on pH and the maximum percent removal 31.4 ± 3.3% has been observed at pH 7. As chitosan was used in the coagulation and flocculation processes, the particles destabilization by chitosan was resulted from charge neutralization and particle entrapment [6]. Due to its cationic behavior, the negatively charged particles in textile suspension are allowed to bind to chitosan *via* ionic or hydrogen bonding [5].

In this study, the adsorption capacity of the dye by chitosan can be improved in the further treatment with either anionic or cationic polymer. After the wastewater samples were treated with 0.1% chitosan for primary coagulation process, the coagulation process of cationic/anionic polymer were performed by using polymer concentrations of 0.1, 0.2 and 0.4 g/l at pH 6-8, respectively. The color removal efficiency of textile wastewater samples before and after treatment with chitosan followed by cationic polymers (CSC) or chitosan followed by anionic polymers (CSA) has been evaluated. Figure 2 indicates the effectiveness of chitosan followed by anionic or cationic polymer to observe the optimal dosage and pH in color removal. It was observed that the color removal efficiency was highest at pH 8 and 7 for CSC and CSA, showing 44.1% and 41.8% removal, respectively. However, there was a statistically significant difference between CSC and CSA as shown in the Figure ($p < 0.05$). %Color removal also related to concentration of ionic polymers as shown in Figure 3. Under the optimum pH (CSC at pH 8 and CSA at pH 7), CSC and CSA indicated the highest color removal at concentration of 0.2 g/l and 0.1 g/l showing 46.5% and 41.8% removal, respectively. Consequently, the optimum parameters for polymer-biopolymer coagulation process in this study was CSC at dosage of 0.2 g/l, pH 8 and CSA at dosage of 0.1 g/l, pH 7. There was a statistically significant difference between CSC and CSA at different concentration as shown. Yadav *et al.* reported that the performance of the coagulation process declines when the pH increases. An increase in alkalinity can alter the nature of hydrolysis products which may influence the destabilization of the colloidal particles. Generally, colloidal particles in water are negatively charged [7]. A change in a coagulant dose may be attributed to the reversal of charge. At higher coagulant concentration, the development of positive charge on the surface of colloidal particles caused by adsorption of excess positively charged ions can promote repulsive interaction and hinder settling.

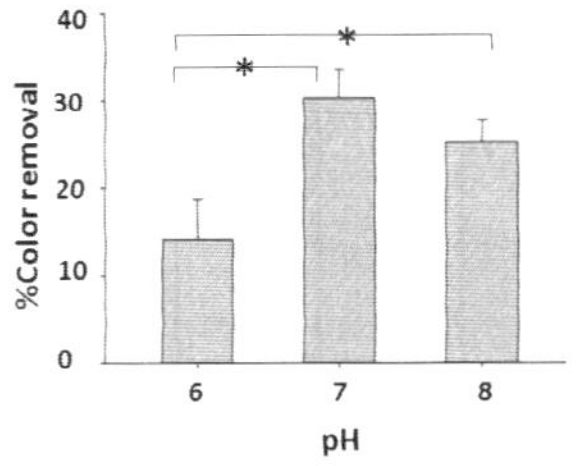

Figure 1. Color removal efficiency after coagulation by 0.1% chitosan at pH 6-8 (*, $p < 0.05$).

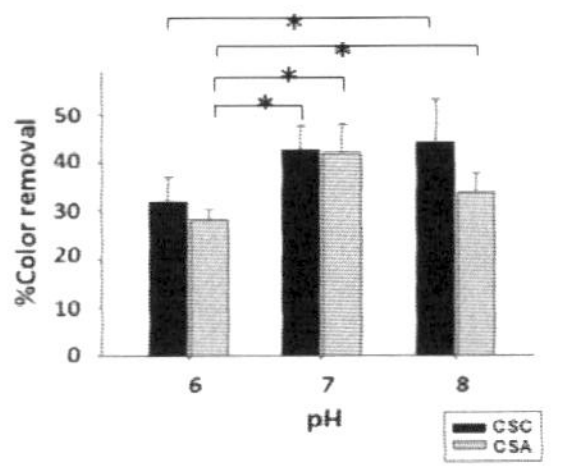

Figure 2. Color removal efficiency after coagulation by CSC and CSA (0.1 g/l ionic polymers) at pH 6-8 (*, $p < 0.05$).

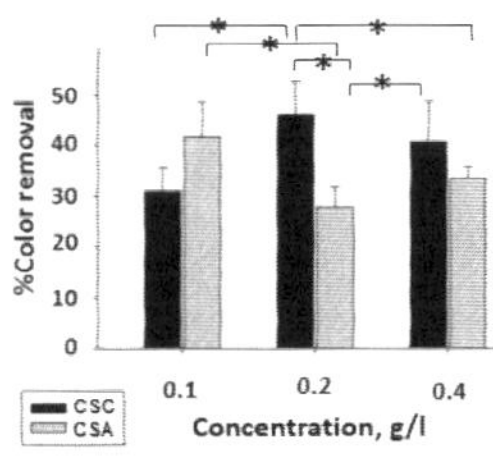

Figure 3. Color removal efficiency after coagulation by CSC at pH 8 and CSA at pH 7 using 0.1-0.4 g/l ionic polymers (*, $p < 0.05$).

Figure 4(A-C) shows the results for COD, DS and SS of the pre-treated wastewater, wastewater after treatment with CSC and wastewater after treatment with CSC. Also, Figure 4D shows the physical characteristics of wastewater before and after treatment at pH 7. . It can be seen that the treated water was obtained with the better quality after treatment with either CSC or CSA in primary coagulation step.

The results indicated that the COD, DS and SS of pre-treated wastewater were 290, 3998.7 and 280.7 mg/l, respectively, however, both DS and SS were higher than the regulatory standard for wastewater. After treatments, all parameters were within the regulatory standard. As the optimum condition, the COD, DS and SS of wastewater after treatment with CSC (0.2 g/l at pH 8) were 260.7, 1939.8 and 103.6 mg/l, respectively, while the COD, DS and SS of wastewater after treatment with CSA (0.1 g/l at pH 7) were 190.6, 3002.2 and 106.8 mg/l, respectively. Interestingly, there was a statistically significant difference between pre-treated and all treated samples (CSC and CSA) at dose of 0.1, 0.2, 0.4 g/l at pH 6-8 ($p < 0.05$). The lower COD, DS and SS is attributed to the higher removal efficiency of coagulants. CSA was effective as coagulant that CSC for COD reduction while CSC was more effective coagulant with respect to the removal performance for DS and SS. Therefore, the combination of chitosan and cationic/anionic polymers could have potential application in textile wastewater treatment.

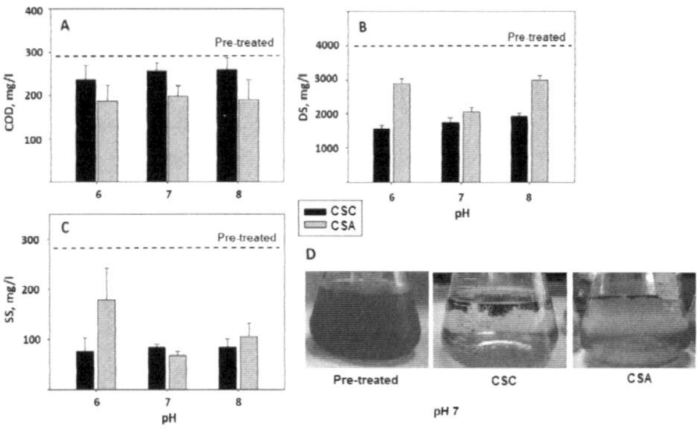

Figure 4. COD, DS, SS and color of wastewater before and after treatment with CSC and CSA. (A) COD of pre-treated wastewater and wastewater after coagulation by CSC and CSA at pH 6-8. (B) DS of pre-treated wastewater and wastewater after coagulation by CSC and CSA at pH 6-8. (C) SS of pre-treated wastewater and wastewater after coagulation by CSC and CSA at pH 6-8. (D) Color of wastewater before and after treatment with CSC and CSA at pH 7.

4. Conclusion

In this study, the wastewater from textile industry can be treated using chitosan together with cationic/anionic polymer via coagulation and flocculation process. CSC exhibited higher efficient COD removal while CSA could achieve higher efficiencies in decreasing DS and SS. Both CSC and CSA were effective for removal of turbidity with value for color removal of 30-47%. The optimum parameters for polymer-biopolymer coagulation process in this study was CSC at dosage of 0.2 g/l, pH 8 and CSA at dosage of 0.1 g/l, pH 7.

5. References

[1] Ghaly A E, Ananthashankar R, Alhattab M V, Ramakrishnan V V. 2014 *J Chem Eng Process Technol.* **5**(1) 1-19.
[2] Rajasulochana P, Preethy V. 2016 *Resource-Efficient Technologies* **2**(4) 175-84.
[3] Rizzo L, Lofrano G, Belgiorno V. 2010 *Separation science and technology* **45**(16):2447-52.
[4] Crini G. 2005 *Progress in polymer science* **30**(1) 38-70.
[5] Hassan MA, Li TP, Noor ZZ. 2009 *Journal of Chemical and Natural Resources Engineering* **4**(1) 43-53.
[6] Desbrieres J, Guibal E. 2018 *Polymer International* **67**(1) 7-14.
[7] Yadav A, Mukherji S, Garg A. 2013 *Industrial & Engineering Chemistry* **52**(30) 10063-71.

Synthesis of sodium lignosulfonate from lignin extracted from oil palm empty fruit bunches by acid/ alkaline treatment for reinforcement in natural rubber composites

N Thungphotrakul[1], P Dittanet[1], S Loykulnunt[2], S Tanpichai[3], P Parpainainar[1]

[1] National Center of Excellence for Petroleum, Petrochemicals and Advance Material, Department of Chemical Engineering, Faculty of Engineering, Kasetsart University, Bangkok 10900, Thailand
[2] Natural Rubber Focus Unit, National Metal and Materials Technology Center, NSTDA, 111 Paholyothin Rd., Klong 1, Klong Luang, Pathumthani 12120 Thailand
[3] Learning Institute and Cellulose and Bio-based Nanomaterials Research Group, King Mongkut's University of Technology Thonburi, Bangkok, 10140, Thailand

* Corresponding author: fengpwn@ku.ac.th

Abstract. This work studied a synthesis of sodium lignosulfonate (SLS) from the lignin which extracted from oil palm empty fruit bunches (OPEFB) by using batch method. The preparation of lignin from OPEFB was done by acid pretreatment using 1% (wt/wt) H_2SO_4 at 90 °C for 120 min in order to reduce hemicelluloses. Then lignin was extracted by 2.5% (wt/wt) NaOH at 90 °C for 180 min and precipitated by adjusting pH to 2 with 98% H_2SO_4. The obtained lignin was changed to SLS by using sodium bisulfite ($NaHSO_3$). The study yields of prepared SLS conducted by varying concentration of $NaHSO_3$. The obtained SLS was characterized by Fourier transform infrared spectroscopy and zeta potential compared to sodium lignosulfonate (commercial).

1.Introduction

Natural rubber is one of important polymers produced in Thailand. Natural rubber has high elastic properties. However, Natural rubber for industrial products has been vulcanized and blended with other biopolymer such as cellulose, lignin and chitosan to improve the properties and sustainable product. In the southern region of Thailand, oil palm industry has an important role in the propel economic in country and that generates large amount of oil palm empty fruit bunches (OPEFB). OPEFB are the biomasses waste from crude palm oil industry and it has molecular structure composed of 30–40 %wt of cellulose, 20–25 %wt hemicellulose and 20–35 %wt of lignin, containing also some quantities of pectin, protein, extractives (nonstructural sugars, nitrogenous material, chlorophyll and waxes) and ash [1]. Base on the study OPEFB, it has about 17.2% of lignin which is high when compared with hardwood and soft wood materials [2]. The isolation of lignin by using direct alkaline treatment is also accompanied by the loss of hemicellulose therefore the biomasses have to be pretreatment with dilute acid to hydrolyze some hemicellulose before alkaline treatment [3]. Lignin is the second most abundant biopolymer next to cellulose. It is a three-dimensional amorphous biopolymer, highly branched. It consists of crosslinked phenylpropanoid monomer structure, such as p-hydroxyphenyl, guaiacyl and syringyl [4]. The major chemical functional group of lignin are dependent on its genetic origin, and applied extraction processes. The different of chemical function group and network structure of lignin are shown to confer special functional properties such as reinforcing effect, biodegradability, UV-absorption and antimicrobial. Lignin was found unsatisfactory for the reinforcement rubber materials because its difficulty of blending and not being dispersed into the rubber. Sodium lignosulfonate (SLS) is kind of surfactants and product from sulfonation process using sodium bisulfite ($NaHSO_3$) with lignin starting materials. It has certain degree of surface activity because its hydrophobic hydrocarbon backbone and hydrophilic sulfonic group[5]. The compatibility of SLS with non-rubber components such as phospholipids and fatty acids on natural rubber particles in natural rubber latex becomes good because ionic sites of SLS interacted with the non-rubber components of the surface of NR particles via Coulomb interaction [6]. In this work, lignin was extracted from oil palm empty fruit bunches by acid hydrolysis pretreatment to reduce some hemicellulose in OPEFB followed by alkaline treatment to produce the lignin. The SLS was reaction between lignin and $NaHSO_3$. The variation of concentration of $NaHSO_3$ react with lignin, the obtained SLS was investigated by yield of sodium lignosulfonate extraction, Fourier-transformed infrared spectroscopy and Zeta potential.

2.Experiment

Materials: Oil palm empty fruit bunches was supplied from Suksomboon oil refinery 109 Moo 4 Hang Sung, Nong Yai District, Chon Buri 20190. Sodium hydroxide (NaOH) was purchased from Ajax Finechem. 98% sulfuric acid (H_2SO_4) was purchased from Merck in Germany. Sodium bisulfite ($NaHSO_3$) was purchased from Himedia laboratories Pvt. ltd. Methanol was purchased from RCI Labscan Limited and sodium lignosulfonate (commercial) was purchased from Tokyo chemical industry co., ltd.

2.1 Acid hydrolysis pretreatment

Acid hydrolysis pretreatment for lignin extraction. The oil palm empty fruit bunches (OPEFB) was dried in oven at 65°C for 48 h and reduced size by mill. The acid hydrolysis of OPEFB were weighed into a flask with volume ratio represents in terms of mass percentage: 10%, 89% and 1 wt.% of EFB, water and conc. sulfuric acid (H_2SO_4) respectively stirred at 90°C for 120 min of reaction time. The acid treated EFB was washed three times with DI water and dried overnight at 65°C.

2.2 Alkaline treatment

Alkaline treatment for lignin extraction. Acid treated OPEFB was added into the mixture 2.5% (w/w) sodium hydroxide (NaOH) solution (solid: liquid ratio; 1: 10, w/w) and then stirred at a temperature of 90°C with a time of 80 min. The result solution contains solid is mainly cellulose and hemicellulose, while liquid fraction is rich lignin (black liquor). The separation of lignin in liquid fraction by acidifying adjusted to pH 2 by adding 98% (w/w) H_2SO_4 gradually. The participated lignin was filtered, washed with water for several times then, dried at 35°C for 8 h.

2.3 Synthesis sodium lignosulfonate from extracted lignin

Synthesis of SLS was reaction between 2 grams of lignin and $NaHSO_3$ solution. The concentration of $NaHSO_3$ were variation of 0.1, 0.3 and 0.5 M, respectively. The condition of process is constant temperature 90°C and constant reaction time 4 h. SLS liquid phase was evaporated at 100°C then mixed with methanol to precipitate an insoluble bisulfite and filtered. The SLS filtered was evaporated to concentrate the SLS then dried at 65°C to constant weight and the SLS obtained to find the yield analyzed determined by mass balance [1].

2.4 Characterization

Fourier transform infrared spectroscopy (FT-IR) analysis. FT-IR analysis was carried out using a Spectrum 400 FT-IR spectrometer. The spectral range was 4000-400 cm-1. Zeta potential was carried out using a Malvern Nano-ZS90 zetasizer after appropriate dilution of solution using water.

3.Results and discussion

3.1 The yield of sodium lignosulfonate extraction

The yield is parameter to determine the SLS produce from lignin with $NaHSO_3$. The result of SLS produce with variation of concentration of $NaHSO_3$ solution show in table 1. By constant the weight of lignin is 2 grams with temperature and reaction time are constant. The results showed the yield tends to increase along with the increase concentration of $NaHSO_3$. The addition of concentration of $NaHSO_3$ provide the more interaction between lignin and $NaHSO_3$ and yielded more SLS.

Table 1. Effect of concentration $NaHSO_3$ to the yield

Concentration of $NaHSO_3$ (M)	Temperature (°C)	Reaction time (hours)	% Yield (w/w)
0.1	90	4	37.71
0.3	90	4	48.73
0.5	90	4	51.87

3.2 Fourier-transformed infrared spectroscopy (FT-IR)

The FT-IR spectroscopy was investigated function group of lignin and sodium lignosulfonate to confirm the lignin was still remain when it through the acid hydrolysis pretreatment and sulfonate group in lignin structure when added the different concentration of $NaHSO_3$ by compare with commercial sodium lignosulfonate. The results from FT-IR image of lignin extraction after through the acid hydrolysis pretreatment is show in figure 1a. Lignin was still remaining in sample. The band nearly in 3400 cm^{-1} are assigned to OH group in aliphatic and aromatic structure. The couple peak between 2920 cm^{-1} and 2850 cm^{-1} are characteristic of C-H bonds of CH_2 and CH_3 of propyl side chain and the peak from 1100 cm^{-1} to 1260 cm^{-1} to represent the syringyl units. The FT-IR images of prepared SLS with variation of concentration of $NaHSO_3$ compared with commercial SLS is show in figure 1b. the bands at 1048 cm^{-1} and 624 cm^{-1} were assigned to adsorption and deformation of sulfonate group, respectively [7].

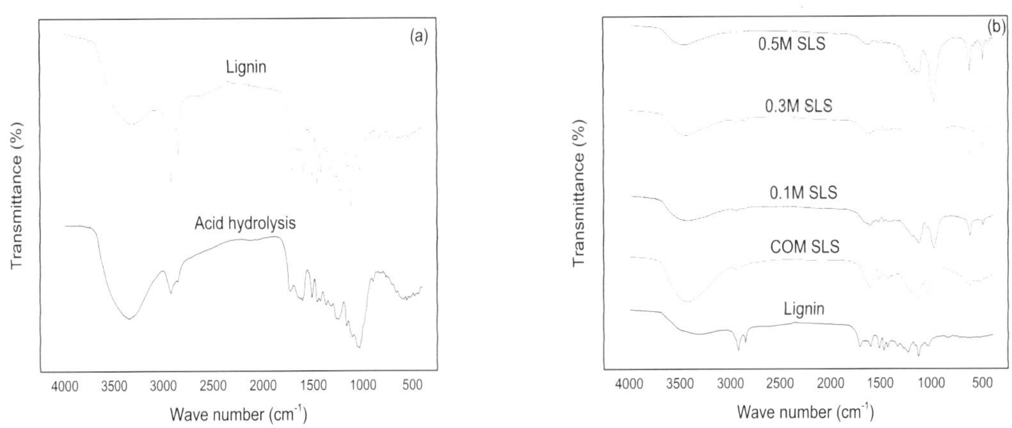

Figure 1. FTIR spectra image of OPEFB after acid hydrolysis and lignin (a) and SLS variation of concentration of $NaHSO_3$ (b).

3.3 Zeta potential

The zeta-potential values, in particular, demonstrate well the electronic charge character of the particles that is inherent to the functional groups of the polymer [8]. The results of lignin and sodium lignosulfonate are negative charged show in table 2. The addition of $NaHSO_3$ in lignin was influence increase the negative charge but when increase the concentration of $NaHSO_3$ influence the trend of negative charge was decrease because may be the high of concentration of $NaHSO_3$ provide the more Na^+ in SLS solution.

Table 2. Zeta potential of lignin and SLS variation of concentration of $NaHSO_3$.

Sample	Zeta potential (mV)
lignin	-53.57
SLS from commercial	-55.17
SLS from 0.1M of $NaHSO_3$	-65.00
SLS from 0.3M of $NaHSO_3$	-60.30
SLS from 0.5M of $NaHSO_3$	-45.47

4. Conclusion

In this work study the synthesis of sodium lignosulfonate from lignin extracted by the sequential acid/alkaline treatment of OPEFB with variation of concentration of $NaHSO_3$ to investigate the trend of yield of synthesis SLS, structure and functional group in synthesis SLS was analyzed by FT-IR and physical distribution of SLS in the solution was analyzed by zeta potential. So that SLS will be use as reinforcing agent and antimicrobial agent in natural rubber composites for further study and maybe it will be use as dispersants to the counter ion for the filler in natural rubber.

5. References

[1] Jesus D C M, Adenise L W, Arion Z F, Lucas B, Miguel D M, Luciana P S V, Sônia F Z and Carlos R S 2016 *Industrial Crops and Products.* **94** 630–637

[2] Jaafar W M K B W, Manan M A, Idris A K and Junin R 2010 *Journal of Materials Science and Engineering* **4** 55-60

[3] Minu K, Jiby K K and Kishore V V N 2012 *biomass and bioenergy* **39** 210-217

[4] Jiang C, He H, Jiang H, Ma L and Jia D M 2013 *eXPRESS Polymer Letters* **7** 480-493

[5] Xinping O, Pan Z, Xueqing Q, Yonghong D and Pu C 2011 *Ind. Eng. Chem. Res.* **50** 10792–10799

[6] Yuko I, Treethip P, Preeyanuch J, Hiroyuki Y, Pranee P, Ritsuko K and Atsushi K 2017 *RSC Adv.* **7** 5222

[7] Nasr A I, Taha M G, Yosef H Y and El-Shaer M A 2017 *Majalah Kulit, Karet, dan Plastik*, **33** 57-64

[8] Suyeon K, Margarida M F, Teresa M, Ana L, Andreia C G and Artur C P 2013 *Colloids and Surfaces B: Biointerfaces* **103** 1-8

Acknowledgement

The authors are grateful that this work was funded by National Center of Excellence for Petroleum, Petrochemicals and Advance Material, Department of Chemical Engineering, Faculty of Engineering, Kasetsart University and NANOTEC Center for Nanoscale Materials Design for Green Nanotechnology.

IOP Conf. Series: Materials Science and Engineering **526** (2019) 012023 doi:10.1088/1757-899X/526/1/012023

Facile preparation of a magnetic carbon adsorbent *via* simultaneous magnetization and activation of sugarcane bagasse and Fe^{2+} and Fe^{3+} ions

S Mahawong[1], A Saning[1], W Watcharin[2], D Dechtrirat[3], and L Chuenchom[1,*]

[1] Department of Chemistry, Faculty of Science, Prince of Songkla University, Hat-Yai, Songkhla 90112, Thailand
[2] Faculty of Biotechnology (Agro-Industry), Assumption University, Hua Mak Campus, Bangkok 10240, Thailand
[3] Department of Material Science, Faculty of Science, Kasetsart University, Bangkok 10900, Thailand

* Corresponding author: laemthong.c@psu.ac.th

Abstract. Magnetic carbon materials have attracted lots of interests as novel adsorbents for removal of toxic substances from aqueous solution. However, so far the preparation of those materials has relied on multi-step procedures, leading to high costs of production in the large scale. Therefore, a facile method to produce magnetic carbon adsorbents *via* simultaneous magnetization and activation process in a single step has been developed. Sugarcane bagasse is the abundantly primary solid waste of sugar industries in Thailand. To best and economically use such sugarcane bagasse, a novel separable magnetic carbon composite (SM-BG-0.2NaOH-800) was successfully prepared *via* the purposed process. Raw sugarcane bagasse was used as a carbon precursor and Fe^{2+} and Fe^{3+} ions as a source of magnetic particles in the carbon composites. NaOH is used as both precipitating agent for the magnetite (Fe_3O_4) particles and activating agent to develop the porosity. The resulting material was characterized using N_2 sorption analysis, X-ray diffraction (XRD), X-ray photoelectron spectroscopy (XPS), Vibrating Sample Magnetometry (VSM).

1. Introduction

Water contaminated with toxic dyes, for example, methylene blue (MB) causes serious harm to humans and ecosystems [1]. The treatment techniques for removing these pollutants include coagulation and flocculation, oxidation, and adsorption [2]. Adsorption is considered to be one of the most effective methods as the result of its simplicity of the design and low costs. Porous carbon materials are a class of excellent adsorbents due to their large specific surface area, developed porosity and chemical stability widely used for dye adsorption in a wastewater system. However, their use in the fine powder form makes the whole process slow due to the sedimentation process. For this reason, magnetic carbon adsorbents have attracted lots of interests owing to their fast and simple separation ability by using an external magnet after completed adsorption [3,4,5].

In this work, the facile single step method for fabricating a magnetic carbon adsorbent *via* simultaneous magnetization and activation process was demonstrated. Sugarcane bagasse is the

IOP Conf. Series: Materials Science and Engineering **526** (2019) 012023 doi:10.1088/1757-899X/526/1/012023

abundantly primary solid waste of sugar industries in Thailand. Therefore, it has been selected as a precursor and Fe^{2+} and Fe^{3+} ions as a source of magnetic particles in the carbon composite. NaOH is used as both precipitating agent for the magnetite iron oxide (Fe_3O_4) particles and activating agent to develop the porosity. Furthermore, methylene blue (MB), a toxic and carcinogenic compound, which has been used in actual textile industry was chosen as a synthetic dyes model for adsorption study.

2. Methodology

First, sugarcane bagasse was collected, washed to remove impurities, and then dried in an oven at 100°C for 48 h. The dried sugarcane bagasse (BG) was ground into 200-325 mesh size. In a typical synthesis, 15.00 g of BG was soaked with 1:2 M ratio of Fe^{2+}/Fe^{3+} for 1 h. Then, 100 mL of distilled water and 38 mL of 0.2 M NaOH solution was added into the mixture subsequently. The resulting mixture was further stirred at ambient temperature for 2 h. Subsequently, it was heated in an oven at 80 °C for 3 days to evaporate the water (SM-BG-0.2NaOH), followed by pyrolysis at 800 °C under nitrogen atmosphere for 1 h. The resulting sample was collected, washed and then labeled "SM-BG-0.2NaOH-800". The magnetic carbon adsorbent was characterized using N_2 sorption analysis, X-ray diffraction (XRD), X-ray photoelectron spectroscopy (XPS), Vibrating Sample Magnetometry (VSM).

To investigate the adsorption performance of the adsorbent towards MB, adsorption isotherm experiments were performed with 0.05 g of the magnetic carbon material and 20 mL of various concentrations of MB solutions. The mixture was shaken for 48 h at room temperature to reach equilibrium. The adsorption kinetics experiments were performed with 0.05 g of the magnetic carbon material and 20 mL of 100 mg L^{-1} of MB solutions, at room temperature (30±2°C) for 0–24 h. After complete adsorption, adsorbent material was separated from the solution using a magnet.

3. Results and Discussion

The XRD patterns of SM-BG-0.2NaOH and SM-BG-0.2NaOH-800 are shown in **Figures 1a,b** respectively. A relatively broad typical XRD pattern for SM-BG-0.2NaOH proves that the uncalcined sample possessed no crystalline iron particles (**Figure 1a**). In contrast, the XRD pattern of SM-BG-0.2NaOH-800 shows sharp peaks at 2θ = 18.3°,30.1°, 35.5°, 37.1°, 43.1°,53.5°, 57.0°, 62.6° and 74.1° which could be indexed to the crystalline magnetite particles (Fe_3O_4) (**Figure 1b**). This could confirm a successful conversion of iron ions into the iron magnetite phase by simple precipitation coupled with a subsequent pyrolysis.

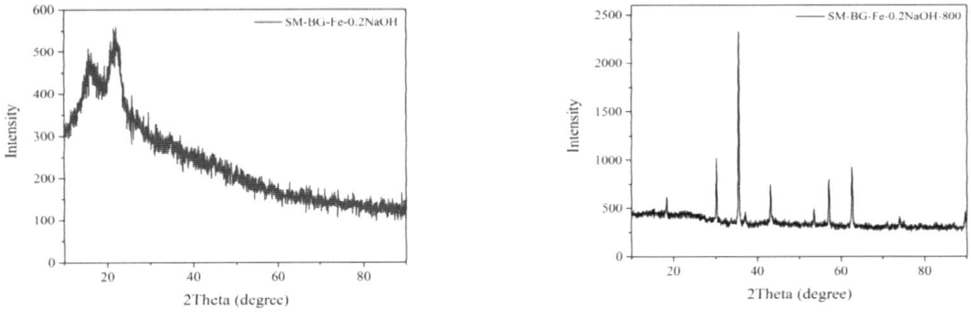

(a) (b)

Figure 1. XRD patterns of SM-BG-0.2NaOH (a) and SM-BG-0.2NaOH-800 (b)

To confirm the development of porosity of the resulting material, N_2 sorption analysis was carried out. The N_2 sorption isotherm of SM-BG-0.2NaOH-800 evidently belongs to type IV isotherm with a small hysteresis loop, according to the IUPAC classification. By calculating using t-plot method, the isotherm is reflective of a large amount of mesopores (65%) and some micropores (25%), respectively. The sample contained the BET surface area of as high as 463 m^2/g and pore volume of $0.23 cm^3/g$

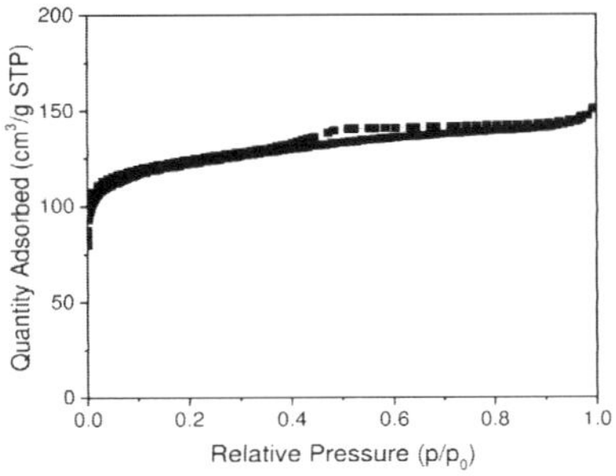

Figure 2. N_2 sorption isotherm of SM-BG-0.2NaOH-800

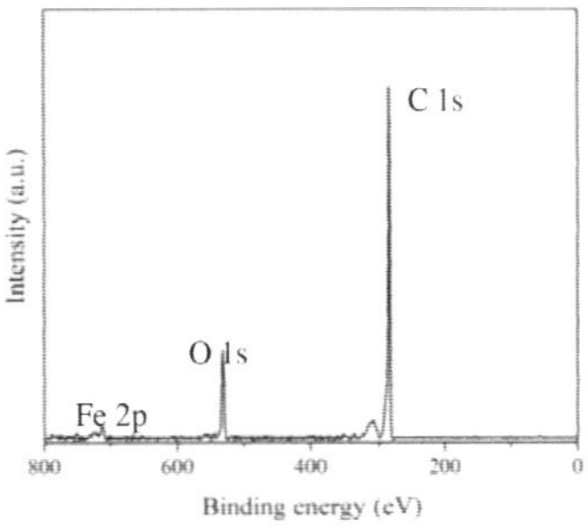

Figure 3. XPS spectrum (survey scan) of SM-BG-0.2NaOH-800

The elemental composition and bonding relationship of SM-BG-0.2NaOH-800 were analyzed by XPS, as shown in **Figure 3**. Three typical peaks corresponding to the binding energies of C 1s (284.5 eV), O 1s (532.0 eV), and Fe 2p (710.38 eV) could be seen in the survey spectra (**Figure 3**), and the Fe 2p peak of SM-BG-NaOH-800 was not very conspicuous due to a low Fe content. XPS analysis determined the C, O and Fe contents to be 89.90, 11.75 and 3.34 %wt, respectively. The high C content and detailed analysis for the XPS (data not shown here) confirms the existence of abundant C sp^2 structure, which is a typical behaviour for the carbons obtained by pyrolysis of biomass at temperature >700°C. Even though the Fe content in the form of Fe_3O_4 for the whole sample (3.34 %wt) is not large, the magnetic properties of SM-BG-NaOH-800 are satisfactory. The good magnetic behavior of the resulting magnetic carbon composite is shown in **Figure 4**. The hysteresis loop evaluated by VSM measurement of SM-BG-0.2NaOH-800 is illustrated in **Figure 4a**. The VSM measurement shows a very high hysteresis and symmetric magnetization curves, suggesting a high super paramagnetic property, which is consistent with the presence of Fe_3O_4 in the XRD results. The saturation magnetization value of the sample was found to be 17.971 emu g^{-1}. As shown in **Figure 4b**, SM-BG-0.2NaOH-800 was easily separated from water by an external magnet and the entire sample particles could completely attracted within 5 minutes.

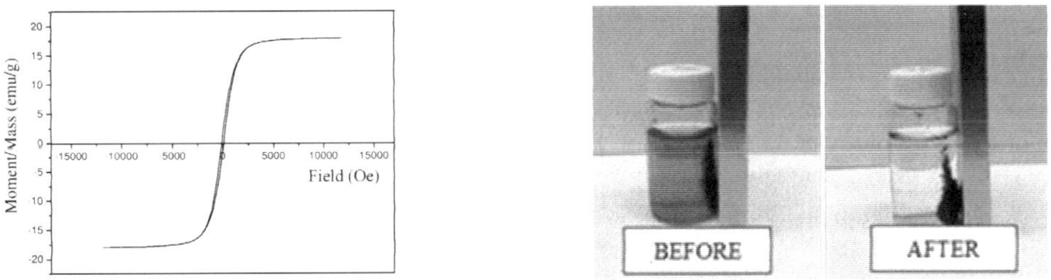

(a) (b)

Figure 4. Magnetization curves obtained by vibrating sample magnetometer (VSM) at room temperature (a) and the magnetic properties before and after the adsorption toward MB dye (20 mg/g, 20 mL) (b) of SM-BG-0.2NaOH-800

The adsorption performance toward methylene blue on SM-BG-0.2NaOH-800 was studies by using 0.05 g of carbon composite in 20 mL of MB solution containing various concentrations for 48 h. The results from adsorption isotherm (**Figure 5a**) demonstrated that the maximum adsorption capacity of MB on SM-BG-0.2NaOH-800 is as high as 58.26 mg/g. The determination coefficient (R^2) of the Langmuir model equal to 1, indicating the isothermal data was well fitted to the Langmuir model, as shown in **Figure 5b**. The adsorption kinetics results are shown in **Figures 6a,b**. In **Figure 6a**, the adsorption amount of MB onto SM-BG-0.2NaOH-800 increased gradually in the first 300 min and reached adsorption equilibrium around 400 min, illustrating a fast adsorption kinetic. The linear fitting curve of pseudo-second-order models for MB adsorption can be observed in **Figure 6b**, exhibiting a good linear correlation with R^2 of 0.9984. The adsorption performance of the sample shown here is

IOP Conf. Series: Materials Science and Engineering **526** (2019) 012023 doi:10.1088/1757-899X/526/1/012023

relatively comparable to or even higher than that of many magnetic adsorbents reported in the literature [5,6,7].

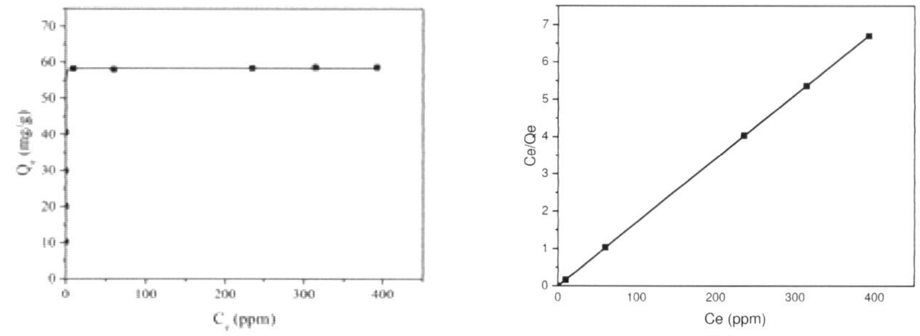

(a) (b)

Figure 5. Adsorption isotherm (a) non-linear and (b) linear plots towards MB dye on SM-BG-0.2NaOH-800.

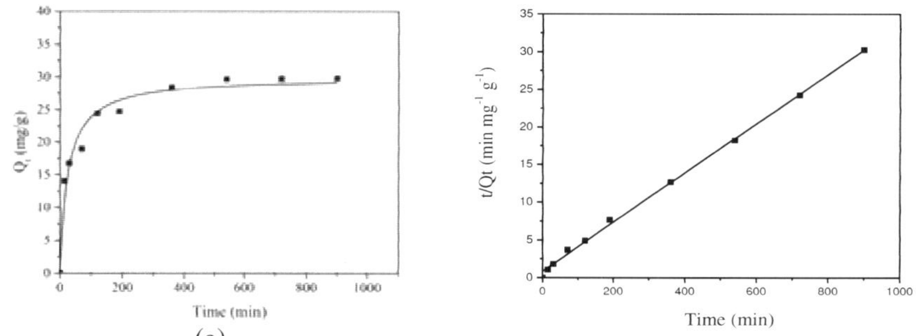

(a) (b)

Figure 6. Adsorption isotherm (a) non-linear and (b) linear plots towards MB dye on SM-BG-0.2NaOH-800.

4. Conclusion

A magnetic carbon adsorbent (SM-BG-0.2NaOH-800) was successfully prepared using sugarcane bagasse as a carbon source and Fe^{2+} and Fe^{3+} ions as a magnetic precursor *via* a facile, one-pot synthesis and simultaneous magnetization/activation process. The adsorbent can be used to adsorb methylene blue dye from aqueous solution, exhibiting adsorption capacity (58 mg g^{-1}) with very high saturation magnetization value (17.971 emu g^{-1}).

5. References

[1] Auerbach S S, Bristol D W, Peckham J C, Travlos G S, Hébert C D and Chhabra R S 2010 *Food and Chemical Toxicology* **48** 169-77
[2] Wan Ngah W S and Hanafiah M A K M 2008 *Bioresource Technology* **99** 3935-48
[3] Rattanachueskul N, Saning A, Kaowphong S, Chumha N and Chuenchom L 2017 *Bioresource Technology* **226** 164-72
[4] Dai J, Tian S, Jiang Y, Chang Z, Xie A, Zhang R, Li C and Yan Y 2018 *Industrial & Engineering Chemistry Research* **57** 3510-22
[5] Kharissova O V, Dias H V R and Kharisov B I 2015 *RSC Advances* **5** 6695-719
[6] Rafatullah M, Sulaiman O, Hashim R and Ahmad A 2010 *Journal of Hazardous Materials* **177** 70-80
[7] Karimifard S and Moghaddam M R A 2018 *Science of the Total Environment* **640** 772-97

Acknowledgements
This work was financially supported by Science Achievement Scholarship of Thailand (SAST), Prince of Songkla University. The Center of Excellence for Innovation in Chemistry (PERCH-CIC), Thailand is acknowledged for partial support.

IOP Conf. Series: Materials Science and Engineering **526** (2019) 012024 doi:10.1088/1757-899X/526/1/012024

Effect of molecular weight of poly(L-lactic acid) on the stereocomplex formation between enantiomeric poly(lactic acid)s blendings

Y Srithep[1,*], D Pholharn[2], and T Akkaprasa[1]

[1]Manufactuing and Materials Research Unit, Faculty of Engineering Mahasarakham University, Mahasarakham 44150, Thailand
[2]Department of Chemistry, Faculty of Science and Technology, RajabhatMahasarakham University, Mahasarakham 44000, Thailand

* Corresponding author: yottha.s@msu.ac.th

Abstract. The effect of molecular weight of poly(L-lactic acid) on the stereocomplex formation between poly(L-lactic acid) (PLLA) and poly(D-lactic acid) (PDLA) was investigated. Films of 1:1 blend were prepared from PLLA with molecular weight of 125, 180 and 209 kg/mol and PDLA with molecular weight of 150 kg/mol by a solution cast method. Thermal properties, crystalline structure and mechanical properties were measured by differential scanning calorimetry (DSC), wide-angle X-ray diffraction (XRD) and tensile testing. Blending lowest molecular weight PLLA (M_w 125 kg/mol) with PDLA showed highest stereocomplex formation with melting temperature more than 50°C compared to that of homocrystallites. As the molecular weight of PLLA became higher, the stereocomplex formation was lower. Films with higher stereocomplex formation also had higher mechanical properties.

1. Introduction

Poly(lactic acid) (PLA) is a plant-derived and compostable polymer with good mechanical, optical and barrier properties [1]. PLA has two enantiomeric isomers, poly(L-lactic acid) (PLLA) and poly(D-lactic acid) (PDLA). For improving thermal/mechanical properties and hydrolysis resistance of poly(lactic acid) (PLA)-based materials, blending PLLA and PDLA leads to the formation of a stereocomplex, which has a melting temperature around 220°C, 50°C higher than homocrystallized PLLA or PDLA [2, 3]. Characteristics of the two components; i.e., the molecular weight, optical purity and chain architecture (linear or branched) have influence over the kinetics and extent of the stereocomplex formation between PLLA and PDLA [4].

In spite of plenty of information available on PLA stereocomplex, in this study, we compared neat and PLA stereocomplex properties with various molecular weights in detail. The stereocomplex formation of PLLA and PDLA was prepared by casting the solutions blend of PDLA and PLLA which had various molecular weights. Stereocomplex formation of the blends was evaluated by differential scanning calorimetry (DSC) and wide-angle X-ray diffraction (XRD). Moreover, the mechanical properties of blended and non-blended films were also evaluated and discussed.

2. Experiments

2.1. Materials
Various molecular weights of PLLA (trade name L105, L130 and L175) and PDLA (trade name D120) was received from Total Corbion PLA (Thailand) Ltd. Table 1 summarizes the properties of poly(lactic acid) used in this work. Molecular weights were measured via a Shimadzu RID-10A GPC (Tokyo, Japan) equipped with a reflective index (RI) detector.

Table 1. Primary data of PLLA and PDLA raw materials

Materials	Trade name	L-unit content (%)	M_n (kg/mol)	M_w (kg/mol)	PDI
	L105	>99	72	125	1.73
PLLA	L130	>99	93	180	1.93
	L175	>99	113	209	1.84
PDLA	D120	<1	89	150	1.69

2.2. Processing
Films used for physical measurements were prepared by solution casting. Briefly, each solution of PLLA and PDLA was separately prepared to have a polymer concentration of 1.0 g/dL and then admixed with each other under vigorous stirring in the case of blended film preparation. Chloroform was used as a solvent and the mixing ratio of PLLA and PDLA was fixed at 1:1. The solutions were cast onto petri-dishes, followed by solvent evaporation at room temperature for approximately 3 days and the obtained sample was further vacuum dried to constant weight. Final film thicknesses were approximately 0.4 mm.

2.3. Material Characterization

2.3.1 Differential Scanning Calorimetry (DSC). Homo and stereocomplex melting temperature (T_{m1} and T_{m2}, respectively) and enthalpy of homo and stereocomplex were determined with a Perkin–Elmer DSC4000 (Massachusetts, USA.) differential scanning calorimeter. Films were heated at the rate of 10°C/min from -10 to 250 °C under a nitrogen gas flow for DSC measurement.

2.3.2 X-ray Diffraction Analysis. X-ray diffraction (XRD) was used to investigate the crystalline structure of the stereocomplex formation (Bruker/D8 Advance BrukerBioSpin AG (Karlsruhe, Germany)). XRD samples were mounted on the XRD platform for analysis. A 2θ range from 2° to 40° in refraction mode was scanned at 2°/min. A computer-controlled wide angle mode goniometer coupled to a sealed tube Cu Kα line source was filtered with a thin Ni filter.

2.3.3 Mechanical Testing. Tensile properties of films were measured at 25°C and 50% relative humidity using a tensile tester (Texture analyser, TA.XT Plus, Stable Micro System, Surrey, UK.) at cross speed of 2 mm/min. Specimens were cut from the cast films with dimension of 1.5 x 4.5 cm.

3. Results and Discussions

3.1. Thermal Properties
Figure 1 shows the DSC thermogram of 1:1 blend of PLLA (L105, L130 and L175) – PDLA (D120) films. The endothermic peaks noticed from 208 to 230°C can be assigned to the melting peak of stereocomplex crystallites and the melting peak around 175°C was the melting peak of homocrystals (Figure 2) [2, 3]. As can be seen from Table 2, melting temperature (T_{m2}) and enthalpy of melting (ΔH_{m2}) of stereocomplex crystallites was highest as the PDLA blended with lowest molecular weight PLLA (L105). As the molecular weight of PLLA increased, T_{m2} and ΔH_{m2} decreased. When PLLA and PDLA are mixed, the lower molecular weight component may diffuse to the higher molecular weight one, resulting in the nucleation and growth of racemic crystallites. On the other hand, a solution of

PDLA and PLLA, both with a high molecular weight, may not be microscopically homogeneous, resulting in preferential homopolymer crystallization [5, 6].

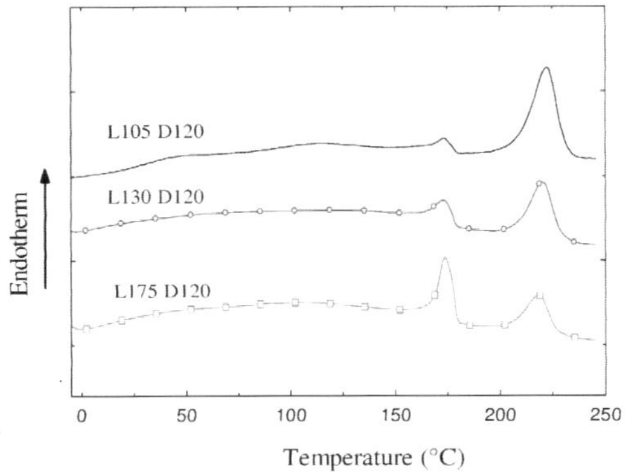

Figure 1. DSC thermogram of blend films of PDLA and PLLA with various molecular weights.

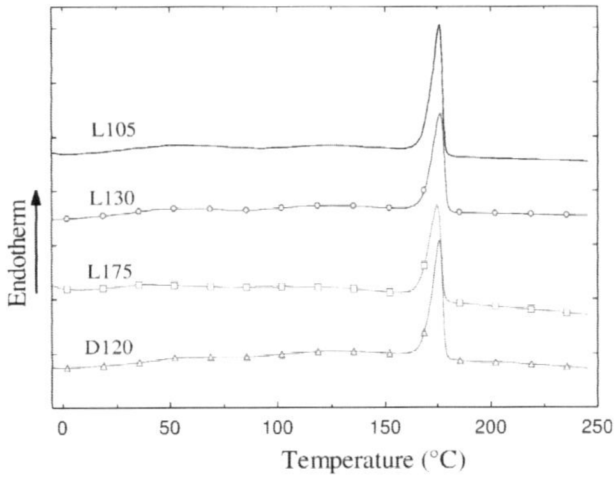

Figure 2. DSC thermogram of neat PLLA and PDLA film.

Table 1. DSC data of blend films of PDLA and PLLA with various molecular weights.

Samples	Melting 1(homocrystal)		Melting 2 (stereocomplex)	
	ΔH_{m1}(J/g)	T_{m1} (°C)	ΔH_{m2} (J/g)	T_{m2} (°C)
L105 D120	3.7	173.5	47.3	221.9
L130 D120	7.3	173.2	26.8	220.3
L175 D120	19.0	173.8	17.8	217.6

3.2. Wide-Angle X-ray Diffraction

Figure 3 shows XRD profiles of blend films of PDLA and PLLA with various molecular weights measured at room temperature (~25°C). The diffraction peaks at 2θ = 11.6°, 20.6°, and 23.5° were assigned to planes of the PLA stereocomplex crystallites (sc) [7-9] and the diffraction peaks located at 2θ about 16.2° and 18.6° corresponded to the α-form homocrystals (Figure 4) [10, 11]. Also, consistent with the DSC result, PDLA (D120) blended with lowest molecular weight PLLA (L105) had highest stereocomplex formation.

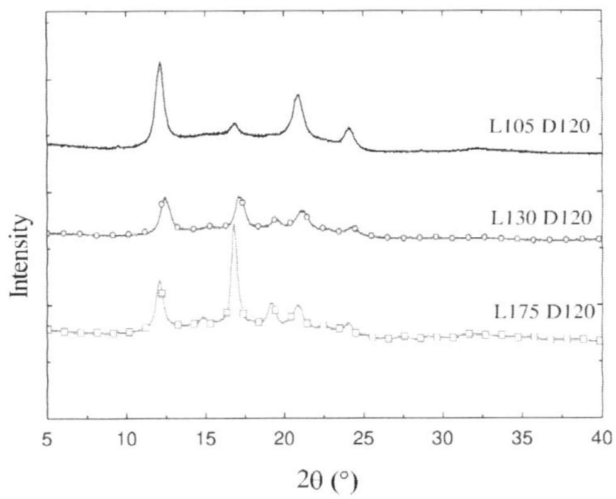

Figure 3. WAXD scans of blend films of PDLA and PLLA with various molecular weights.

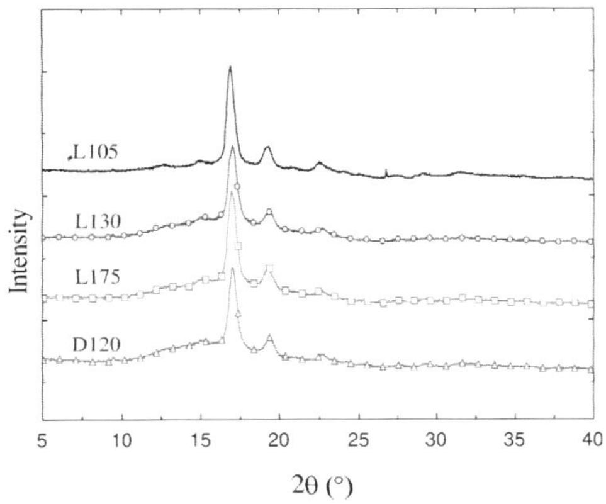

Figure 4. WAXD scans of neat PLLA and PDLA film.

3.3. Tensile Properties

Figure 5 shows stress-strain curves of PDLA and PLLA with various molecular weights. As evident from Figure 1 and 3, stereocomplex crystallites are the main crystalline species in L105 D120. Although L175 had highest molecular weight and mechanical properties (Figure 6), the tensile strength and Young's modulus of the blend L175 D120 were lowest. This indicated that increased tensile strength and modulus properties are observed for the blend films containing higher stereocomplex crystallites. It may be reasonable to assume that a higher density of tie chain for stereocomplex samples will result in better mechanical properties [5].

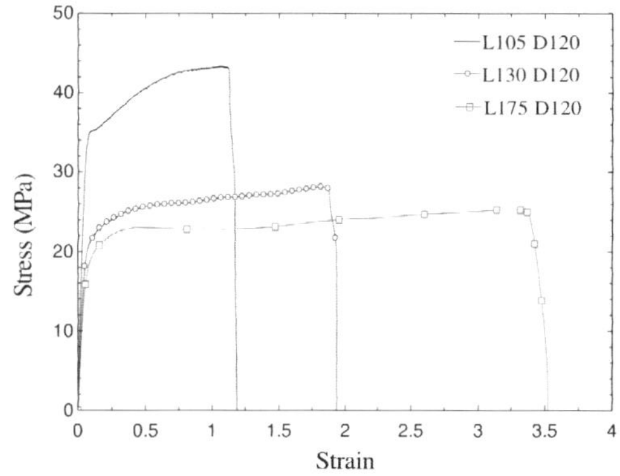

Figure 5. stress-strain curves for PLLA/PDLA blend films.

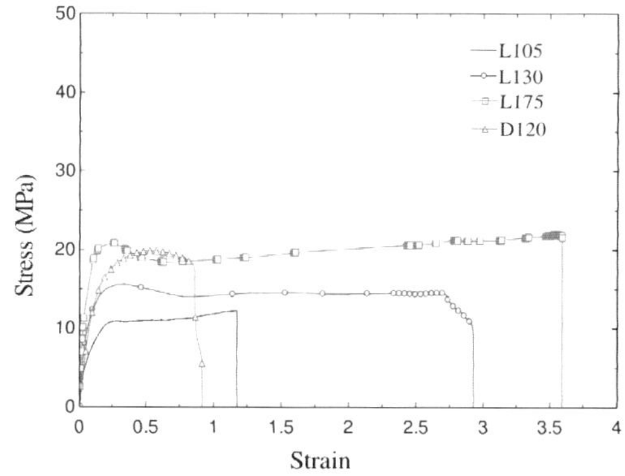

Figure 6. stress-strain curves for neat PLLA and PDLA films.

4. Conclusions

In this article, the PDLA and PLLA with various molecular weights were blended by solution cast method. The blend of PDLA and PLLA which had lower molecular weight formed higher stereocomplex crystallites than that having higher molecular weight. Blend films which are rich in stereocomplex crystallites have better tensile strength and modulus than films rich in homo-crystallites.

5. References

[1] Saeidlou S, Huneault M A, Li H, Sammut P, and Park C B 2012 *Polymer* **53** 5816-24.
[2] Srithep Y, Pholharn D, Turng L-S, and Veang-in O 2015 *Polym. Degrad. Stab.* **120** 290-9.
[3] Liu Y, Sun J, Bian X, Feng L, Xiang S, Sun B, Chen Z, Li G, and Chen X 2013 *Polym. Degrad. Stab.* **98** 844-52.
[4] Saeidlou S, Huneault M A, Li H, and Park C B 2013 *Polymer* **54** 5762-70.
[5] Tsuji H and Ikada Y 1999 *Polymer* **40** 6699-708.
[6] Pan P, Han L, Bao J, Xie Q, Shan G, and Bao Y 2015 *J. Phys. Chem. B* **119** 6462-70.
[7] Bao R-Y, Yang W, Jiang W-R, Liu Z-Y, Xie B-H, Yang M-B, and Fu Q 2012 *Polymer* **53** 5449-54.
[8] Cartier L, Okihara T, and Lotz B 1997 *Macromolecules* **30** 6313-22.
[9] Srithep Y, Nealey P, and Turng L S 2013 *Polym. Eng. Sci.* **53** 580-8.
[10] Wu T-M and Wu C-Y 2006 *Polym. Degrad. Stab.* **91** 2198-204.
[11] Mathew A P, Oksman K, and Sain M 2006 *J. Appl. Polym. Sci.* **101** 300-10.

Acknowledgement
Financial support from the Thailand Research Fund and Office of the Higher Education Commission (MRG6180115) and Faculty of Engineering Mahasarakham University is gratefully acknowledged.

IOP Conf. Series: Materials Science and Engineering **526** (2019) 012025 doi:10.1088/1757-899X/526/1/012025

Biodegradation of Levan Polymer / Poly (Lactic Acid) (PLA) Blend

P Phengnoi[1], K Thongmee[1], S Tiptipakorn[1], B Boekfa[1] and K Kuttiyawong[1,*]

[1] Department of Chemistry, Faculty of Liberal Arts and Science, Kasetsart University, Kamphaeng Saen Campus, Nakhon Pathom 73140, Thailand

*Corresponding author: faaskmt@ku.ac.th

Abstract. Poly (lactic acid) (PLA) is a biodegradable thermoplastic. With low elongation and poor degradation properties, many approaches have been developed to overcome these limitations, including copolymerization, blending and plasticizing to improve the mechanical properties. Levans are a group of fructan polymer with act as a good co-polymer, which can be produced by levansucrase using sucrose as substrate. In this study, co-solvent for levan/PLA blend film preparation, thermal properties and biodegradation of levan/PLA blend film were investigated. The result found that DMSO and CH_2Cl_2 were suitable for the co-solvent process. The Levan/PLA blend films were prepared with levan contents of 0, 0.5, 1, 2, 5, 10, 15 and 20 wt%. The biodegradation of levan/PLA blend film *in vitro* and in soil conditions showed that 15 wt% levan/PLA at 7 days and 10 wt% levan/PLA at 21 days were the highest weight loss, respectively. In addition, the morphological changes in the Levan/PLA blend film and thermal properties were analyzed by Scanning electron microscope (SEM) and Differential scanning calorimetry (DSC). The results from DSC present that the melting temperature of PLA was slightly decreased with increasing levan contents. Moreover, the holes from degradation were larger and the rougher surfaces were observed with levan contents.

1. Introduction

In recent years, there has been using biodegradable polymers from renewable resources in many fields such as medicine, agriculture, and packaging industry [1]. With environmentally friendly materials, they have been interesting to develop morphological, thermal and biodegradability properties. Poly (lactic acid) (PLA) is a biodegradable thermoplastic polymer, which can be obtained from 100% natural sources such as corn, tapioca, as well as from petrochemical resources [2]. PLA doesn't have good properties, especially low elongation, poor hydrophilicity, complicated processes for preparing high molecular-weight polymer and high price [3-6]. Copolymerization, blending and plasticizing methods can be helping to improve these poor properties of PLA [7-8]. Levans are composed of D-fructose units joined together by a β-(2, 6) glycosidic linkage in the main chain with β-(2, 1) linkage at branch points, which can be produced from plants and microorganisms. A superior thermal resistance with the high melting temperature (223°C) and non-toxicity, it has been using in many applications such as food, medicine and nanotechnology [9-12].

In this work presented herein focuses on a co-solvent for levan/PLA blend film preparation, thermal properties and biodegradation of levan/PLA blend film. Biodegradation was studied *in vitro* and in soil conditions. Morphology and thermal properties were determined by Scanning electron microscope (SEM) and Differential scanning calorimetry (DSC).

2. Experimental

2.1. Materials
Sucrose was used as a substrate for the levan production, purchased from Ajex Finechem Pty Ltd., Australia. *E. coli* BL-21 was used as an expression host and pETLSKK9 containing levansucrase gene (lsKK9) of *B. licheniformis* KK9 for levansucrase production. Glucose was used for levansucrase inducing, purchased from APS Finechem, Australia. IPTG was used as a promotor for levansucrase expression, purchased from Sigma-ALORICH, Switzerland. PLA was obtained from BC polymeers marketing, Co., Ltd. Luria-bertani broth contain peptone, yeast extract and sodium chloride, purchased from Himedia Laboratories, BD, USA and Sigma-ALORICH, Switzerland, respectively. Other reagents were used to co-solvent for Lev/PLA blend production.

2.2. Levansucrase Activity Assay
The activity of levansucrase was measured by quantifying the reducing ability of free glucose from the reaction by the modified DNS method [13]. One unit was defined as the amount of levansucrase that produces 1 μmole of reducing sugar (glucose equivalent) per min.

2.3. Levan production
The reaction mixture containing 8.75 % w/v sucrose, 3 U/mL levansucrase in 25 mM citrate buffer pH 6.0 and incubated at 37 ºC, 150 rpm for 16-18 h. after that the levan was precipitated in acetone. The levan was filtered, dialyzed in water and lyophilized.

2.4. Levan/poly (lactic acid) blend preparation.
Solvents for levan and poly (lactic acid) dissolving were studied using DI water, acetone, dichloromethane and dimethyl sulfoxide. The optimum solvents for levan and PLA dissolving were used as co-solvent for Levan/poly (lactic acid) blend preparation. Polymer blend of levan and PLA were prepared. Levan/poly (lactic acid) blend were prepared at various levan/PLA mass ratios (0, 0.5, 1, 2, 5, 10, 15 and 20 wt%) using co-solvent for dissolving. Then place in the compression mold machine at 50 °C until the levan/poly (lactic acid) blend was molded. After that, cut the sample size 1x1 cm to study the biodegradation.

2.5. Biodegradation of Polymers via in vitro condition.
The polymer blend was degraded with *Penicillium sp.* and *Pseudomonas aeruginosa*. Preparation of *Penicillium sp.* and *Pseudomonas aeruginosa* for the biodegradation were cultured in medium containing 0.25 g yeast extract, 0.5 g $(NH_4)_2SO_4$, 0.5 g KH_2PO_4, 0.8 g K_2HPO_4, 0.1 g $MgSO_4 \cdot 7H_2O$, 0.05 g NaCl, 0.01g $CaCl_2 \cdot 2H_2O$, 0.005 g $FeSO_4 \cdot 7H_2O$, 0.25 mg $Na_2MoO_4 \cdot 2H_2O$ and 0.25 mg $MnSO_4$ in 500 mL DI water at 37 ºC, 250 rpm for 16-18 h. Dropping 5 μL of 5 mL culture medium on polymer blend samples in tubes and incubated at 37 ºC, 250 rpm for 3, 5, 7, 14 and 21 days. The polymer blend samples were measured weight loss and captured surface.

2.6. Biodegradation of Polymers: in soil condition.
The polymer blend was placed and filled soil in a porous plastic box. Embedding and watering a porous plastic box 1 time per day for 21 days. The polymer blend samples were measured weight loss and captured surface.

3. Results and discussion

3.1. Optimum solvents for the levan / poly (lactic acid) blend preparation

Co-solvent for the levan / poly (lactic acid) blend was investigated. Dimethyl sulfoxide (DMSO) and dichloromethane (CH_2Cl_2) showed a great levan/poly (lactic acid) dissolving property. These results indicate that optimum dissolution of levan in DMSO was 11.4 % and PLA in CH_2Cl_2 was 20% w/v. Further, both types of solvent mix very well and homogeneously.

3.2. Levan /poly (lactic acid) blend preparation

The mass ratios of levan/PLA (0, 0.5, 1, 2, 5, 10, 15 and 20 wt%) were investigated. At 0 wt% levan, the PLA polymer was transparent but other mass ratios (0.5, 1, 2, 5, 10, 15 and 20 wt%) showed that the levan/PLA blends were not transparent.

3.3. Biodegradation via in vitro condition

The results from the biodegradation revealed that the holes from biodegradation were observed after 3 days on 15 and 20 wt% levan in PLA. While the other sample ratios were observed the holes in several days. Except the pure PLA sample, the surface did not biodegrade in 21 days. %Weight loss showed that at 0, 0.5, 1 and 2 wt% levan in PLA had the weight loss in range 0 – 15 wt% in 21 days, which compared with 10, 15 and 20 wt% levan in PLA had more increasing trend in range 25 – 60 wt% in 7 days and decreasing trend in range 20 – 50 wt% after 7 days (Figure 1). The highest weight loss of levan/PLA was 56.55 % at 15 wt% levan in 7 days biodegradation test However, the weight loss of levan/PLA blend via *in vitro* condition may cause be swelling and hydrolysis from water in solution.

3.4. Biodegradation in soil condition

The results from the biodegradation showed that the holes from biodegradation were clearly observed on 10, 15 and 20 wt% levan in PLA. While the other sample ratios were observed the slightly holes. %Weight loss indicated that at 0, 0.5, 1 and 2 wt% levan in PLA had the weight loss in range 0 – 8 wt% in 21 days, which compared with 10, 15 and 20 wt% levan in PLA had more increasing trend in range 10 – 45 wt% in 21 days (Figure 2). The highest weight loss of levan/PLA was 41.86 % at 10 wt% levan in 21 days of biodegradation test. Previously report, the biodegradation rate of P(3HB, 4HB)/PLA have highest increasing in 5 months [14]. These results indicated that the levan contents act as a key to biodegradability of PLA via *in vitro* and soil conditions.

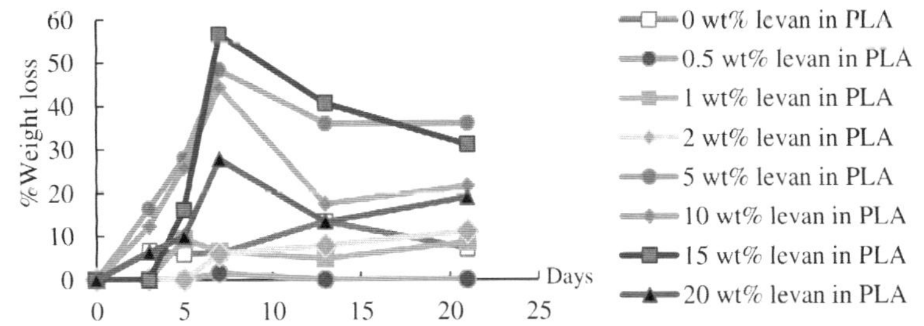

Figure 1. %Weight loss of levan/PLA blend from biodegradation via in vitro condition

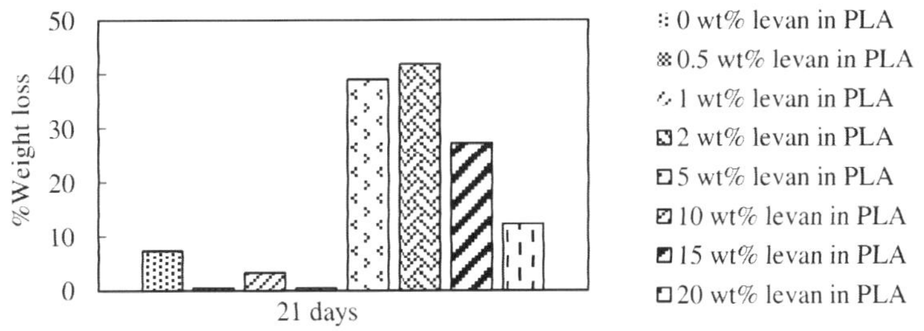

Figure 2. %Weight loss of levan/PLA blend from biodegradation in soil condition for 21 days

3.5. Effects of levan on thermal properties of levan/PLA blend

The results from Differential Scanning Calorimetry (DSC) presented that the melting temperature of levan/PLA blend was slightly decreased from 154.8 to 154.5 °C. This mean there was no significant effect of levan on melting behavior. But the levan content showed the area under a graph of melting point (Tm) less than the PLA without levan. This result indicated that the levan/PLA blend had less crystalline and more amorphous than PLA without levan. According to crystallization temperature (Tc), which showed that the levan/PLA blend had slightly decreased crystalline when prepared with the PLA without levan, resulting in an easy biodegradation.

4. Conclusion

The biodegradation of levan/PLA blend was investigated. It was found that newly prepared polymer blend for the fast biodegradation in both *in vitro* and soil conditions. The biodegradation showed that the holes from biodegradation were clearly observed. At 15 wt% levan in PLA had highest degradation was 56.55 % via *in vitro* condition in 7 days. In part of soil condition, at 10 wt% levan in PLA had highest degradation was 41.86 % in 21 days. DSC measurements revealed the melting temperature of levan/PLA blend was slightly decreased from 154.8 to 154.5 °C. However, the levan/PLA blend still need to be improved to obtain efficiently in application.

5. References

[1] Uhrich K E, Cannizzaro S M, Langer R S and Shakesheff K M 1999 *Chem. Rev.* **99** 3181-98.
[2] Gupta A P and Kumar V 2007 *Eur. Polym. J.* **43** (10) 4053-74.DOI:10.1016 /j.eurpolymj.2007.06.045.
[3] Lunt J 1998 *Polym. Degrad. Stab.* **1-3** (3):145-52. DOI: 10.1016/S0141-3910(97)00148-1.
[4] Rasal R M, Janorkarc A V and Hirta D E 2010 *Prog. Polym. Sci.* **35** (3) 338–356. DOI: 10.1016/j.progpolymsci.2009.12.003.
[5] Drumright R E, Gruber P R and Henton D E 2000 *Adv. Mater.* **12** (23) 1841–46. DOI: 10.1002/1521-4095(200012)12:23<1841::AIDADMA1841>3.0.CO;2-E.
[6] Lima L T, Auras R and Rubino M 2008 *Prog. Polym. Sci.* **33** (8) 820–52. DOI: 10.1016/j.progpolymsci.2008.05.004.
[7] Kucharczyk P, Sedlarik V, Miskolczi N, Sakacs H and Kitano T 2012 *J. Reinf. Plast. Compos.* **31** (3) 189-202. DOI: 10.1177/0731684411434150.
[8] Martin O and Avérous L 2001 *Polymer.* 2001 **42** (14) 6209–19. DOI: 10.1016/S0032 3861(01)00086-6.
[9] Vuyst L De, Vin F De, Vaningelgem F and Degeest B 2001 *International Dairy Journal.* **11** (9) 687-707.
[10] Manatsittipan S, Kuttiyawong K and Tiptipakorn S 2017 *Key Eng. Mater.* **737** 294-8.
[11] Kang S A, Jang K H, Seo J W, Kim K H, Kim Y H, Rairakhwada D, Seo M Y, J Lee O, Ha S and Kim C 2009 *Caister Academic Press*, Norwich. 145-61.
[12] Liu Z, Jiao Y, Wang Y, Zhou C and Zhang Z 2008 *Advanced Drug Delivery Reviews.* **60** (15), 1650-62.
[13] Miller G L 1959 *Anal. Chem.* **31** 426-8.DOI 10.1021/ac60147a030.
[14] Weng Y, Wang L, Zhang M, Wang X, and Wang Y 2013 *Polymer testing.* **32** 60-70.

Acknowledgements

Financial support for this research was granted through a scholarship of the Graduate School, Kasetsart University, Year 2016, to Ponsakorn Phengnoi. The authors would like to acknowledge the Center for Advanced Studies in Nanotechnology for Chemical, Food and Agricultural Industries, KU Institute for Advanced Studies. And also, would like to acknowledge the Thailand Research Fund (TRF): MRG5480232.

IOP Conf. Series: Materials Science and Engineering **526** (2019) 012026 doi:10.1088/1757-899X/526/1/012026

Optimizationof poly(L-lactide)-polybutylene adipate terephthalate diblockcopolymer by ring opening polymerization

D Pholharn[1] and Y Srithep[2,*]

[1] Department of Chemistry, Faculty of Science and Technology, Rajabhat Maha Sarakham University, Mahasarakham 44000, Thailand
[2] Manufactuing and Materials Research Unit, Faculty of Engineering Mahasarakham University, Mahasarakham 44150, Thailand

* Corresponding author: yottha.s@msu.ac.th

Abstract. Polylactide (PLA) is a brittle biodegradable polymer: co-bonding of a higher flexibility polymer such as polybutylene adipate terephthalate (PBAT) into the PLA chains has been shown to make it more flexible. We studied the optimum conditions for synthesizing poly(L-lactide)-polybutylene adipate terephthalate diblock copolymer (PLLA-*b*-PBAT), with ratios by weight of L-lactide:PBAT from 100:0.3 to 100:1. Incubation times of polymerization were varied from 4 to 8 hours at 160°C. Nuclear magnetic resonance (NMR) spectra were used to verify the structure of synthesized polymers. Differential scanning calorimetry thermograms showed melting temperatures of diblock copolymers at around 161-168°C - characteristic of PLA. The optimum conditions for PLLA-*b*-PBAT synthesis by ring opening polymerization were 100:0.3 ratio and eight hour incubation leading to 95% yield, a high degree of crystallinity (71%) and the highest molecular weight (34,000g/mol, measured by gel permeation chromatography).

1. Introduction

Nowadays, there is considerable interest in polylactide (PLA) from industrial, medical and research fields because it has many advantages such as biodegradability, easy processing, nontoxicity and biocompatibility[1]. There is special interest in one of its optical isomers, poly(L-lactide)(PLLA)[2]. However, brittleness, poor thermal stability and low crystallization rate present significant disadvantages of PLA [3]. Blending PLA with other more flexible biodegradable polymers such as poly(butylenes adipate terephthalate)(PBAT)would result in higher elongation at break of PLA [4].Although PBAT and PLA have a similar carbonyl groups on their backbone chains, low interfacial adhesion and macro-phase separation between them lead to immiscibility. Studies have reported that the addition of more than 5% of PBAT into PLA could cause phase separation [5]. In order to reduce that problem, this study tried to co-bond between PLLA and PBAT to form a single chain of diblock copolymer via bulk ring opening polymerization from L-lactide. The reactions were optimized by varied ratios of L-lactide:PBAT and incubation time when the temperature was constant at 160°C.

2. Experimentals

2.1 Materials
98% L-lactide(LL) was donated by Purac (Thailand) Ltd. For ring opening polymerization, a stannous octoate ($Sn(Oct)_2$) used as catalyst was supplied by Sigma Aldrich and PBAT(ecoflex®; M_n ~ 100,000 g/mol) acted as self macro-initiator and was obtained from BASF corporation. Solvents, including toluene, methanol, tetrahydrofuran (THF) and chloroform, were obtained from RCI Lab scan.

2.2 Synthesis of poly(L-lactide)-polybutylene adipate terephthalate diblock copolymers
Poly(L-lactide)-polybutylene adipate terephthalate diblock copolymers (PLLA-*b*-PBAT) were prepared via bulk ring opening polymerization of L-lactide. Before using, LL was dried in a vacuum oven at 45°C for 12h. The ring opening polymerization proceeded in nitrogen atmosphere at 160°C. It initiated by PBAT that have a hydroxyl group on chain-end. PBAT would act as maro-initiator by itself and reacted with LL under insertion-coordination mechanism [6]. The ratios by weight of LL:PBAT were 100:0.3 100:0.6 and 100:1. $Sn(Oct)_2$ 0.1mol% was used as a catalyst in the reaction. The reaction times was 4 and 8 h. After incubation, the product was dissolved in chloroform and removed to purify by precipitation in methanol. The precipitate was filtered under vacuum. Finally, PLLA-*b*-PBAT was dried in a vacuum oven at 50°C for 24 h. The synthesis scheme of the PLLA-*b*-PBAT by ring opening polymerization is shown in Figure 1.

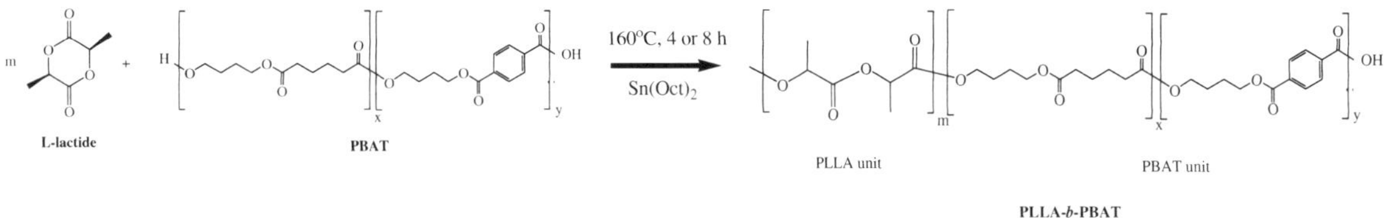

Figure 1. Synthesis of PLLA-*b*-PBAT by ring opening polymerization

2.3 Measurements and characterization

2.3.1 Yield percent. The yield percent of product obtained from the reactions was calculated by equation (1)

$$\text{Yield percent } (\%) = \frac{W_{DB}}{W_{LL} + W_{PBAT}} \times 100 \tag{1}$$

Where W_{DB} is weight of dried PLLA-*b*-PBAT, W_{LL} is weight of L-lactide and W_{PBAT} is weight of PBAT before incubation in the reaction.

2.3.2 Gel permeation chromatography (GPC). The M_n, M_w and PDI of PLLA-*b*-PBATs were measured by GPC.GPC columns were packed with Styragel HR5E 7.8 × 300 mm columns. A Waters 2414 refractive index (RI) detector was used. The GPC columns were eluted using THF at a flow rate of 1.0 mL/min at 40°C and monodispersed polystyrenes were used for calibration.

2.3.3 Differential scanning calorimetry (DSC). Thermal properties of the samples were analyzed by DSC (Perkin-Elmer, model DSC 4000). Approximately 5 mg of samples were placed in aluminium pans and subjected to two heating scans from -10 to 200°C at 10°C/min. The melting temperature (T_m) and apparent melting enthalpy (ΔH_m) were determined from the second heating scan. The degree of crystallinity $\chi_c(\%)$ was estimated from the first heating cycle using

$$\chi_c(\%) = \frac{\Delta H_m}{\Delta H_m^o} \times 100 \tag{2}$$

where ΔH_m is the heat of fusion of PLA and $\Delta H_m^o$is theoretical melting enthalpy of perfect PLA crystal(93 J/g) [7]

2.3.4 Proton Nuclear Magnetic Resonance (^{1}H-NMR). To verify the chemical structure of PLLA-*b*-PBAT, a Bruker Advanced DPX 300 MHz ^{1}H-NMR spectrometer was used. $CDCl_3$ was used as a solvent at ambient temperature and tetramethyl silane was used as internal chemical shift standard.

3. Results and Discussion

3.1 Percent yield and molecular weight
Yield percent and molecular weight of PLLA-*b*-PBATs are presented in Table 1. It is remarkable that when the incubation time was increased from 4 to 8 h, the yield percent of PLLA-*b*-PBATs increased around 10%. The higher yield percent indicated a more complete reaction. Using a 8h incubation time, we found that different ratios of LL:PBAT influenced M_w and M_n of the synthesized polymers. Highest molecular weight of PLLA-*b*-PBAT (M_w = 34,000 g/mol and M_n = 22,000 g/mol) was obtained when the LL:PBAT was 100:0.3.These results indicate that the most efficient condition for ring opening polymerization was 8hincubation time and LL:PBAT was 100:0.3.

3.2 Chemical structure
The NMR spectra of PLLA, PBAT and PLLA-*b*-PBAT are shown in Figure 2.The signals at 5.16 (peak a) and 1.56 (peak b) correspond to the methine proton (-OCH) and methyl proton (-OCH_3) in the PLLA unit [8-10].The PBAT segment has two polymer chain; chain A and B (see structure in Figure 2). At 2.34 ppm (peak c) assigned to outer CH_2 groups in adipic acid, 4.11-4,17 ppm (peak e) was the outer CH_2 groups in 1,4-butanediol unit in chain A . At 4.34-4.41 ppm (peak f) assigned to the outer CH_2 groups of 1,4-butanediol unit and 8.11 ppm (peak g) meant to CH in benzene ring of chain B. Further, the signal found at 3.67-3.75 ppm (peak d), assigned to the CH_2 groups of the 1,4-butanediol adjacent to the terminal –OH group of chain A in PBAT[5]. However, it entirely disappeared in spectrum of PLLA-*b*-PBAT. This clearly demonstrated that the hydroxyl group of PBAT initiated the L-lactide in the ring opening polymerization.

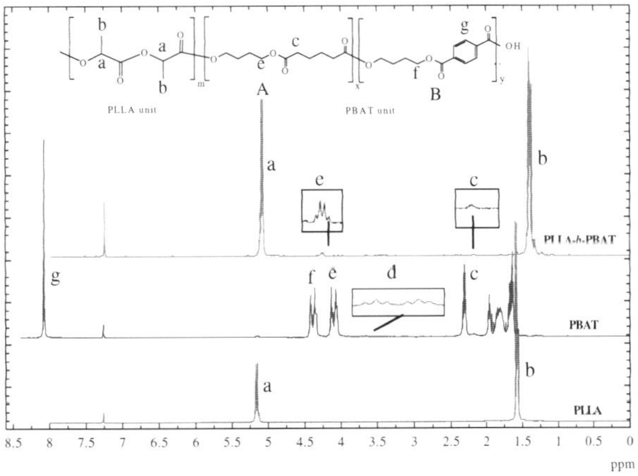

Figure 2.NMR spectra of PLLA, PBAT and PLLA-*b*-PBAT

3.3 Thermal properties
The thermal properties of the samples were analyzed by DSC. DSC thermogram and the data are shown in Figure 3 and Table 1, respectively. The melting peak of all PLLA-*b*-PBAT samples occurred at around 161-168°C that is the characteristic of PLA [11].It is obvious that when using LL:PBAT at 100:0.3 and incubated for 8 h, the degree of crystallinity of PLLA showed a high value (70.89%). These results correlate with the NMR spectrum and confirm that the synthesized polymer were actually possessed PLA structure.

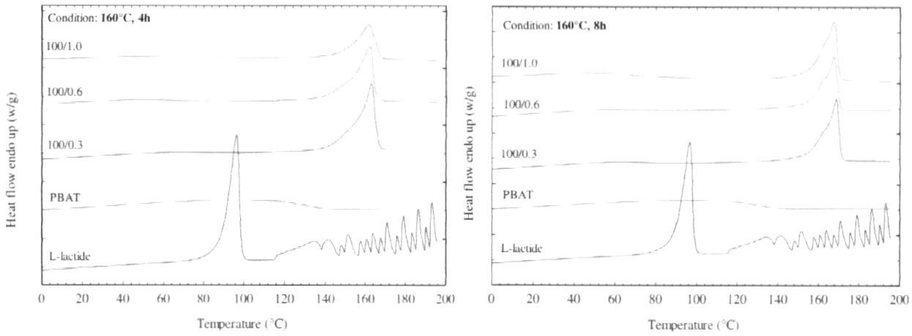

Figure 3. DSC thermogram of L-lactide, PBAT and PLLA-*b*-PBATs with different condition and LL:PBAT ratio

Table 1. Yield percent, molecular weight and DSC data of PLLA-*b*-PBATs

Samples	Temp. (°C)	Time (hour)	M_w (g/mol)	M_n (g/mol)	PDI	Yield percent	ΔH_m (J/g)	T_m (°C)	χ_c (%)
100:0.3			7,600	6,400	1.19	83	74.01	162.90	79.58
100:0.6		4	17,900	9,300	1.92	87	64.55	162.00	69.41
100:1.0			16,800	11,000	1.52	82	41.91	161.65	44.09
100:0.3	160		34,000	22,000	1.56	95	65.93	168.61	70.89
100:0.6		8	31,500	18,600	1.69	96	56.42	167.47	60.67
100:1.0			26,700	13,200	2.03	93	65.86	167.38	70.82
PBAT	-	-	-	-	-	-	93.28	95.55	-
LL	-	-	-	-	-	-	91.85	96.45	-

4. Conclusions

The PLLA-*b*-PBAT was successfully synthesized via bulk ring opening polymerization of L-lactide and PBAT acting as macro-initiator. The optimum condition found from this study was using ratio of LL:PBAT at 100:3 and incubating the reaction for 8 h. This conclusion is inferred from the highest of yield percent (95%) and molecular weight (M_w = 34,000 g/mol). The NMR and DSC results could mutually verify the structure of PLLA-*b*-PBAT. Future work of this study will be mechanical properties and heat resistance of PLA triblock copolymer.

5. References

[1] Yang B, Raza S, Li S, and Deng J 2017 *Polymer* **127** 214-9.
[2] Lopes M S and Jardini A L 2014 *Chem. Eng. Trans.*
[3] Jing Z, Shi X, Zhang G, Li J, Li J, Zhou L, and Zhang H 2016 *Polymer* **92** 210-21.
[4] Freitas A L P d L, Tonini Filho L R, Calvão P S, and de Souza A M C 2017 *Polym. Test.* **62** 189-95.
[5] Ding Y, Lu B, Wang P, Wang G, and Ji J 2018 *Polym. Degrad. Stab.* **147** 41-8.
[6] Ryner M, Stridsberg K, Albertsson A-C, von Schenck H, and Svensson M 2001 *Macromolecules* **34** 3877-81.
[7] Xiao H, Lu W, and Yeh J T 2009 *J. Appl. Polym. Sci.* **113** 112-21.
[8] Kaitian X, Kozluca A, DENKBAŞ E B, and PİŞKİN E 1996 *Turk. J. Chem.* **20** 43-53.
[9] Supthanyakul R, Kaabbuathong N, and Chirachanchai S 2017 *Polym. Degrad. Stab.* **142** 160-8.
[10] Ba C, Yang J, Hao Q, Liu X, and Cao A 2003 *Biomacromolecules* **4** 1827-34.
[11] Jing Z, Shi X, Zhang G, Li J, Li J, Zhou L, and Zhang H 2016 *Polymer* **92** 210-21.

Acknowledgements

We acknowledge financial support from the Research and Development Institute, Rajabhat Maha Sarakham University and Mahasarakham university for suporting the instrumemts.

IOP Conf. Series: Materials Science and Engineering **526** (2019) 012027 doi:10.1088/1757-899X/526/1/012027

Efficacy of cold atmospheric pressure plasma jet against *Enterococcus faecalis* in apical canal of human single-rooted teeth: a preliminary study

K Saleewong[1], P Wanachantararak[2], P Louwakul[1]

[1]Department of Restorative Dentistry and Periodontology, Faculty of Dentistry, Chiang Mai University, Chiang Mai, 50200, Thailand
[2]Dental Research Center, Faculty of Dentistry, Chiang Mai University, Chiang Mai, 50200, Thailand

* Corresponding author: kanruethai_sa@cmu.ac.th

Abstract. Many factors that may affect the success of root canal treatment are complexity of root canal system, especially in the apical part, and the virulence of bacteria such as biofilm formation. A cold atmospheric pressure plasma jet is a novel method for sterilization and applied for root canal disinfection. The purpose of this study was to investigate the effect of the cold atmospheric pressure plasma jet on *Enterococcus faecalis (E. faecalis)* biofilms infected in the apical root canals. **Methods:** Fifty-four single rooted-teeth were infected with *E. faecalis* for seven days and allocated into five groups. Group I performed as positive control. Group II-V were served as experimental groups: NaOCl, plasma, NaOCl + plasma and gas only, respectively. The disinfection was evaluated by colony count (per milligram) at the depths of 0.05 and 0.10 mm of root dentin. The remaining bacteria was also counted from the grounded root dentin. Data was analysed by one-way ANOVA and post hoc test, with a level of significance set at $P<.05$. **Results:** The NaOCl, plasma and NaOCl + plasma groups significantly reduced the *E. faecalis*, but no significant difference was found among these groups ($P>0.5$). The NaOCl + plasma group significantly reduced *E. faecalis* in the deeper dentin level compared to the other groups. **Conclusion:** The cold atmospheric pressure plasma jet had antimicrobial activity against *E. faecalis* biofilm as well as the use of NaOCl. The combined treatment yielded the most promising result and may be beneficial in root canal disinfection.

1. Introduction

The goals of root canal treatment are irritants removal, cleaning and obturation of shaped root canal, and preventing the root canal re-infection. Even though, the conventional root canal treatment is 40-60%[(1)] reduction of total amount of living bacteria due to not only the complexities of root canal anatomy especially in the apical parts of the roots such as isthmi, apical ramifications and apical deltas, but also microbial in the root canals always form as the biofilms that compose of extracellular polysaccharide (EPS) protecting microbes. The most commonly isolated bacterial species in retreatment cases with apical pathoses are *Enterococcus faecalis (E. faecalis)* which is able to form the biofilm, attach to the dentinal tubules in the root canals, survive from medicaments i.e. Calcium hydroxide; $Ca(OH)_2$ [(2-5)].

There are several advanced development methods which anticipating for efficiently cleaning the root for example LASERS, nanoparticle, and plasma. Recently, the plasma, a partially ionized gas produced by applying an electric or electromagnetic field to one or more gases[6, 7], is introduced to the biomedical filed including medicine and dentistry. For dentistry, the plasma was studied in many fields such as dental caries, increase bond strength, or root canal disinfection[8]. In this in-vitro study, we evaluated the effectiveness of cold atmospheric plasma jet as a disinfection method for the *E. faecalis* biofilm in the apical root canal while different levels of root canal dentin depth were investigated.

2. Materials and Methods

2.1 Preparation of apical root canals

The methods of this study were previously approved by the Human Experimentation Committee, Chiang Mai University, Thailand. Fifty-four single rooted-teeth were cleaned and stored in 0.1% thymol solution before the experiment. The teeth were horizontally cut to acquire the apical root length of 10 mm using diamond discs. Dental pulps were removed. A15 K-file was placed into the root canal and progressed through the canal until it was just visible at the major apical foramen. The working length was recorded as one mm short of this length. The coronal part of root canals was prepared by gates-glidden drill number 2 and 3 at 6- and 8-mm (from the anatomical root apices), respectively. The remaining root canals were enlarged by using nickel-titanium files (Race: FKG Dentaire SA, La Chaux-de-Fonds, Switzerland) to size 40.02. The 5% sodium hypochlorite (NaOCl) was used as an irrigant during the preparation. Each specimen was subjected to final irrigation steps as followed (five minutes each): immersing in 17% EDTA in an ultrasonic bath to remove smear layer, cleaning with 5% NaOCl in an ultrasonic bath and soaking in 3.84% sodium thiosulfate solution to neutralize effects of NaOCl. The apical foramina were sealed with resin composite (Meta Biomed, Seoul, South Korea). Every specimen was submerged into a 50-ml centrifugal tube containing 20 mL of Brain Heart Infusion (BHI) broth. Sterilized all teeth in an autoclave at 121°C for 15 minutes.

2.2 Enterococcus faecalis biofilm formation

E. faecalis (ATCC 29212) was incubated in BHI broth at 37°C for three hours. The concentration of bacteria was 1×10^8 cells/mL measured by an optical density at 600 nm. For the biofilm formation, each canal was filled by 20 µL of BHI broth containing E. faecalis and aerobically incubated at 37°C for seven days. Media in the root canals was changed every two days.

2.3 Cold atmospheric pressure plasma device

A commercial bioplasma jet (Photo Bio Care Co. Ltd, Nonthaburi, Thailand) which employed in skin clinic was used as a power supply for this experiment (Figure 1). The plasma was generated along tungsten wire (length of 1.2 cm and diameter of 0.25 mm) acting as an electrode and 1.7-cm ethylene-tetrafluoroethylene (ETFE) needle (Terumo Surflo®: Terumo Corporation, Tokyo, Japan) as a dielectric barrier. Working gases were helium and oxygen (98% He and 2% O_2). The flow rate of 0.5 L/min was adjusted by Cole Palmer flow rate meter. The frequency and intensity were set at 10 kHz and 100 Hz, respectively.

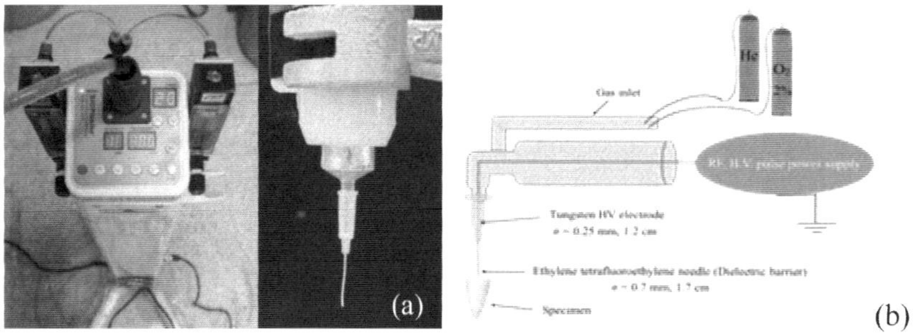

Figure 1. A photograph of experimental plasma setting (a) and its schematic (b).

2.4 E. faecalis biofilm disinfection and assessments

The specimens were randomly divided into controls (N=18) and four experimental groups (N=36). The control groups did not acquire any treatment and were incubated at 37°C for seven days. In the positive control group, the roots were only infected with *E. faecalis* and there was only BHI broth in the negative control group. For the experimental groups (group 1-4): (1) five mL of NaOCl for one minute, (2) five-minute cold plasma treatment, (3) five mL of NaOCl for 1 minute with 5-minute cold plasma treatment and (4) gas only (He/O_2). The plasma tip was placed at 5 mm from the working length. After exposure of each treatment, the root canal dentin was collected by two size files of Race (50.02 and 60.02) into the pre-weighed 1.5-mL Eppendorf tube. The remaining root was freeze in -70 °C liquid nitrogen and was pulverized. The weighed of each dentin level was determined. The 10-fold serial dilutions were generated and incubated at 37°C for 24 hours. The bacterial concentration was calculated as CFU/mg. The difference between the group was analyzed by one-way ANOVA and post hoc test. $P < .05$ was considered as statistical difference.

3. Result

Mean $\log_{10}$ CFU data at the level of 50/.02 and 60/.02 for different experimental groups showed that the NaOCl, plasma, and combined groups were effective for reducing the bacterial number ($P < .05$) and these treatments were more effective than the positive control group and treatment with only working gas ($P < .05$). Moreover, the combined group inclined more elimination of *E. faecalis* biofilm in the deeper level of root canal dentin (Figure 2).

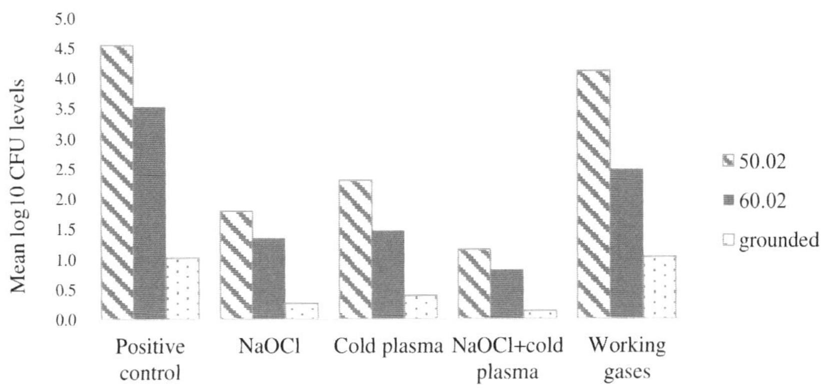

Figure 2. Mean log10 CFU levels in infected root canals after disinfection treatments in different root canal depths.

4. Discussion

Apical periodontitis usually results from an infection in the root canals which bacteria always formed as the biofilms. The biofilm is more resistant to antimicrobial substances than planktonic bacteria(4). The *E. faecalis* species is commonly chosen as developed models for testing *in vitro* study designed for endodontic disinfection since it was frequently isolated from the persistent periapical diseases(2-4). Success rate of the conventional root canal treatment is about 68-85%(9). The utilization of cold plasma treatment in Endodontics is to enhance the elimination of root canal bacteria. The reactive species that play an important role in sterilization are low level UV radiation and RONS (reactive oxygen and nitrogen species). The RONS have oxidative effects on the bacterial cell walls so the structure integrity is disturbed(10). Meanwhile, an atmospheric non-thermal plasma or cold plasma as it not only reacts to the bacterial cell membrane leading to ruptured cells and finally cell death, but also causes no or little heat damage to both direct and indirect cell contact(11).

In this study, samples were divided into four experimental groups: NaOCl, cold plasma, NaOCl + plasma and gas only. For the positive control group, *E. faecalis* formed biofilms attaching to root canal surfaces and some penetrated into dentinal tubules more than a depth of 600 μm. The NaOCl is a widely used irrigant in bacterial disinfection and dissolving tissues during endodontic treatment. The NaOCl groups demonstrated decreasing the number of *E. faecalis* in every depths of root canal dentin. The cold plasma group showed significant bacterial reduction as well as in the NaOCl and NaOCl + plasma group. The combination of NaOCl and cold plasma showed the most bacterial reduction, especially in deeper dentin levels. The cold plasma in the gas phase has ability penetrating into deeply root canal system, so the plasma may be able to reach the residue bacteria after using the NaOCl. [12, 13] To increase the success rate of root canal treatment, maximum bacterial elimination before obturation is a key factor[14]. Therefore, the combination of cold plasma with conventional root canal treatment may be used as adjunctive.

For the gas group, the amount of *E. faecalis* was not significantly different from the positive control group indicating that the He and O_2 alone were not ineffective in killing bacteria without the impulse of electric or electromagnetic fields to break the gas molecules into ions according to previous studies[15].

5. Conclusion
The cold atmospheric pressure plasma jet had antimicrobial activity against *E. faecalis* biofilm as well as the use of NaOCl. In addition, the combined treatment yielded the most promising result and may be beneficial in root canal therapy.

6. References
[1]. Sjogren U, Figdor D, Persson S, Sundqvist G. Influence of infection at the time of root filling on the outcome of endodontic treatment of teeth with apical periodontitis. *Int Endod J. 1997 30(5) 297-306..*
[2]. Molander A, Reit C, Dahlén G, Kvist T. Microbiological status of root-filled teeth with apical periodontitis. *Int Endod J.*. 1998 **31(1)** 1-7.
[3]. Sundqvist G, Figdor D, Persson S, Sjogren U. Microbiologic analysis of teeth with failed endodontic treatment and the outcome of conservative re-treatment. Oral Surg Oral Med Oral Pathol Oral Radiol Endod. 1998 **85(1)** 86-93.
[4]. Kishen A, Haapasalo M. Biofilm models and methods of biofilm assessment. *Endodontic topic Topics*. 2010 **22** 58–78.
[5]. Love RM. Enterococcus faecalis--a mechanism for its role in endodontic failure. *Int Endod J.* 2001 **34(5)** 399-405.
[6]. Ureyen Kaya B, Kececi AD, Guldas HE, Cetin ES, Ozturk T, Oksuz L, et al. Efficacy of endodontic applications of ozone and low-temperature atmospheric pressure plasma on root canals infected with Enterococcus faecalis. *Letters in applied microbiology*. 2014 **58(1)** 8-15.
[7]. Kim JH, Lee MA, Han GJ, Cho BH. Plasma in dentistry: A review of basic concepts and applications in dentistry. *Acta Odontologica Scandinavica*. 2013(1-12).
[8]. Cha S, Park YS. Plasma in dentistry. *Clin Plasma Med*. 2014 **2(1)** 4-10.
[9]. Ng YL, Mann V, Gulabivala K. A prospective study of the factors affecting outcomes of nonsurgical root canal treatment: part 1: periapical health. *Int Endod J*. 2011 **44(7)** 583-609.
[10]. Laroussi M. Nonthermal decontamination of biological media by atmospheric-pressure plasmas: review, analysis, and prospects. *IEEE Transactions on Plasma Science*. 2002 **30(4)** 1409-15.
[11]. Fridman G, Friedman G, Gutsol A, Shekhter AB, Vasilets VN, Fridman A. Applied Plasma Medicine. *Plasma Process Polym*. 2008 **5** 503-33.
[12]. Habib M, Hottel TL, Hong L. Antimicrobial effects of non-thermal atmospheric plasma as a novel root canal disinfectant. *Clin Plasma Med*. 2014;2(1):17-21.
[13]. Ritts AC, Li H, Yu Q, Xu C, Yao X, Hong L, et al. Dentin surface treatment using a non-thermal argon plasma brush for interfacial bonding improvement in composite restoration. *European journal of oral sciences*. 2010 **118(5)** 510-6.
[14]. Siqueira JF, Jr., Rocas IN. Clinical implications and microbiology of bacterial persistence after treatment procedures. *J Endod*. 2008 **34(11)** 1291-301.
[15]. Du T, Ma J, Yang P, Xiong Z, Lu X, Cao Y. Evaluation of Antibacterial Effects by Atmospheric Pressure Nonequilibrium Plasmas against Enterococcus faecalis Biofilms In Vitro. *J Endod*. 2012 **38(4)** 545-9.

Acknowledgement
The authors are grateful to the Department of Restorative Dentistry and Periodontology, Faculty of Dentistry, Chiang Mai University for the financial support.

IOP Conf. Series: Materials Science and Engineering **526** (2019) 012028 doi:10.1088/1757-899X/526/1/012028

Numerical Investigations on the Effect of Gas Flow Rate in the Gas Stirred Ladle with Dual Plugs

S Lakkum[1], **P Kowitwarangkul**[1*]

[1] The Sirindhorn International Thai-German Graduate School of Engineering (TGGS), King Mongkut's University of Technology North Bangkok (KMUTNB), Bangkok, 10800, Thailand

E-mail: pruet.k@tggs.kmutnb.ac.th

Abstract. Ladle or secondary steelmaking is an essential metallurgy process in which the steel is adjusted to the chemical compositions required in the final product. The objective of this study was to predict 95% mixing time curve using CFD simulation in order to investigate the possibility to optimize the gas flow rate in gas stirred ladle systems. The commercial software, Flow-3D, was used in this study. This study was divided into three parts. The first two parts of the study was the investigation of the effects of bubble sizes and diffusion coefficient on the mixing time by setting constant-volume gas flow rate. The last part was the study of the effect of gas flow rate on the mixing time. The ladle geometry and operation condition from the example steel plant, Millcon Steel PLC were used in this study. The simulation results indicated that different sizes of bubble have minor effects on the mixing time. The investigation of the effect on the mixing time with and without diffusion coefficient showed that the diffusion coefficient has no significant influence on the mixing time but has significant effects on the characteristics of the plume regions and the velocity flow field. Finally, it was found that gas flow rate has significant effects on turbulent kinetic energy, the mixing time and the steel cleanliness. The results can support the example steel plant to optimize the mixing process concerning productivity and cleanliness quality of the liquid steel.

1. Introduction

The stirring with an inert gas is widely used in the ladle refining process [1]. In the refining process, the gas (mostly argon) is injected into the liquid steel through porous plugs [2]. The gas bubbles from porous plugs generate the recirculation flow pattern in the ladle, which enhances mixing efficiency to homogenize the chemical composition and temperature [1]. The desired degree of mixing used is normally fixed at 95%, which means that the mixing time is the time when compositions reaches or remains constant within 5% range of the final concentration [3]. The mixing time is an important factor that determines the operation time and the mixing efficiency in the ladle furnace. Steelmaking plants often reduce the operation time by using high gas flow rate to increase level of productivity. However, excessive gas flow rates or long stirring times may cause the slag entrainment and re-oxidation [4,5]. Thus, it is necessary to optimize the gas flow rate that maintains the mixing efficiency in the ladle furnace in order to reduce the inclusion in the liquid steel.

Over the past decades, there has been a number of research studies into the flow and mixing phenomena in gas stirred ladle systems. D. Mazumdar et. al. [6] studied the physical modeling and mathematical modeling of gas stirred ladle systems. In their study, a number of literature reviews on various phenomena of gas stirred ladle systems such as turbulent fluid flow, mixing, gas-liquid interactions, etc. has been investigated. The results provided the understanding of the extensive mathematical modeling which lead to the process simulations. Numerous previous studies [7-8] presented the use of Froude scaling criterion. It was shown that the appropriate range of gas flow rates in the two systems between the model and the full-scaled ladle are related according to $Q_{mod} = \lambda^{2.5}Q_{f.s}$. The research works [1,2,9] described the bubble size prediction model which is related to the stirring flow rate.

The purpose of this study is to investigate the effects of the bubble sizes, the diffusion coefficient and the gas flow rate on the mixing time in a gas stirred ladle with dual plugs configurations by using numerical simulation, the Flow-3D software. The ladle geometry and operation condition from the example steel plant, Millcon Steel PLC are used in this numerical study. A 1:5 scaled water model is also numerically simulated in order to compare with the experimental results of water model in the next study. The results from this study can support the example steel plant to optimize the gas flow rate concerning operation time and cleanliness quality of the liquid steel.

2. Geometry of ladle model and simulation conditions

A 1:5 scaled water model of the example steel plant was used in the simulation. Figure 1 shows the ladle furnace of the example steel plant, Millcon Steel PLC (a), the arrangement of porous plugs and the tracer inlet (b) and meshing model of the ladle (c). The simulation was carried out via the discrete phase model (DPM), which was used to track motions of individual bubble, volume of fluid model (VOF) and RNG k-ε turbulence model. The total number of the cells used for solving flow equation is 734,456 cells. The dimensions of the 1:5 scaled water model and the full-scaled system are shown in Table 1.

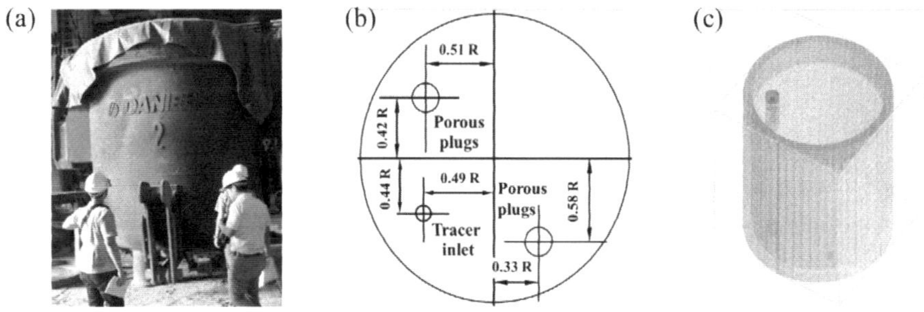

Figure 1. Geometric model of the ladle: (a) The full-scaled ladle furnace, (b) Arrangement of porous plugs and the tracer inlet and (c) Meshing model

Table 1. The simulation conditions and ladle dimensions

Model	Liquid density (kg/m^3)	Liquid viscosity (Pa.s)	Gas Density (kg/m^3)	Liquid temperature (K)	Ladle diameter (mm)	Porous plugs diameter (mm)	Ladle height (mm)	Liquid height (mm)	Tracer cylinder height (mm)	Tracer cylinder diameter (mm)
Water model scale 1:5 (Water-Air)	1000	0.001	1.225	298	467	20.4	600	500	120	17
Full-scale (Steel-Argon)	7020	0.0055	1.6228	1873	2335	102	3000	2500	600	85

The scope of this study was divided into three parts. The first two parts were carried out using a 1:5 scaled model. The constant-volume gas flow rate for both cases is 5.4 L/min. The first part was studied by adjusting different sizes of the bubble between 2 mm and 5 mm with a constant diffusion coefficient. The second part was investigated by setting the simulations with and without diffusion coefficient with bubble sizes of 2 mm and 5 mm. For the last part, the simulation of both water model and the full-scaled (steel-argon) ladle furnace were carried out using the geometrically and dynamically similar systems. The gas flow rates in full-scaled system were reduced from 300 L/min,

200 L/min and 100 L/min to 5.4 L/min, 3.6 L/min and 1.8 L/min, respectively. Additional simulation conditions are shown in Table 1.

For the boundary conditions, no-slip boundary condition was used at the bottom and side wall. The gas bubbles were assumed to escape at the surface of liquid. Tracer was injected for two seconds at tracer inlet with fixed mass fraction and flow rates. At the porous plug, the gas flow rate and the bubble diameter were specified.

3. Simulation results

3.1 The effects of bubble sizes and diffusion coefficient on the mixing time

The examples of 95% mixing time curve of the 2 mm and 5 mm bubbles are shown in Figure 2(c). The results from the simulation indicated that the mixing time of the 5 mm bubble is around 220 seconds which is 15 seconds faster than that of the 2 mm bubble. It can be seen that the different sizes of bubbles have minor effects on the mixing time. The results of diffusion coefficient on the mixing time indicated that the mixing time with diffusion coefficient is 10 seconds faster than the mixing time without diffusion coefficient. Cone-shaped plumes occur in the simulation with diffusion coefficient as shown in Figure 2(a, b), while in case without diffusion coefficient, the plume regions appear in columnar shapes. The diffusion coefficient has no significant effect on the mixing time but has significant influence on the characteristics of the plume regions and the velocity flow field as shown in Figure 2(a, b).

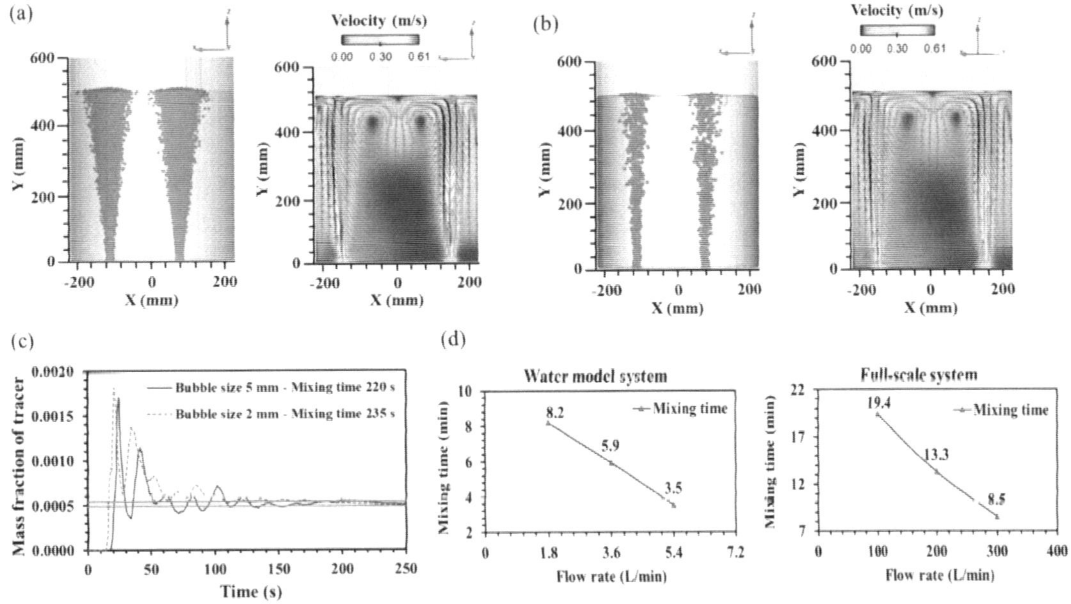

Figure 2. The results from the simulation; Plume regions and the velocity flow field of the bubbles with diffusion coefficient (a) 2 mm, (b) 5 mm, (c) 95% mixing time curve comparing between 2 mm and 5 mm bubble, (d) The mixing time from water model and full-scaled (steel-argon) systems

3.2 The effect of gas flow rate on the mixing time and the turbulent kinetic energy

The simulation results showed that high gas flow rate decreases the mixing time in the 1:5 scaled water model and full-scaled systems as shown in Figure 2(d). In addition, high gas flow rate significantly increases the turbulent kinetic energy as shown in Figure 3. Typically, high turbulent kinetic energy increases slag open-eye size which results in re-oxidation and slag entrainment on the liquid surface. Therefore, the gas flow rate should be optimized in order to improve the cleanliness quality of the liquid steel by decreasing the turbulent kinetic energy.

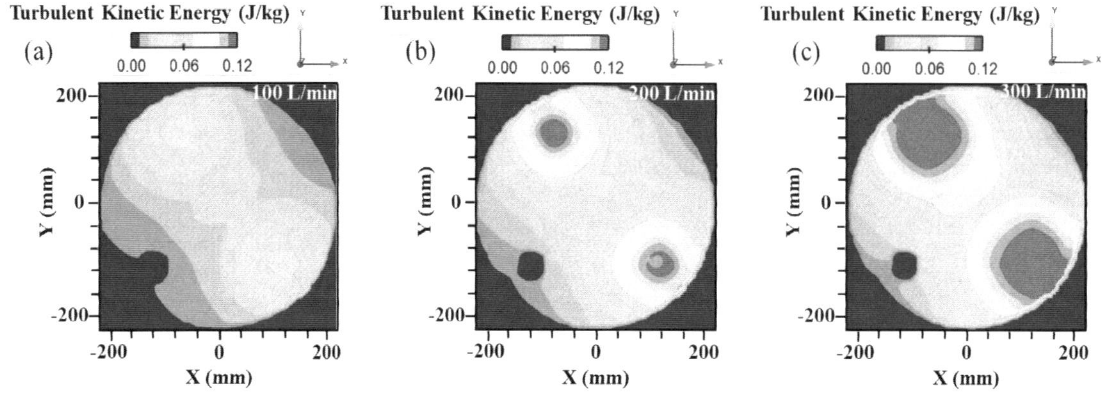

Figure 3. The turbulent kinetic energy with different flow rates from a full-scaled (steel-argon) system

4. Conclusion

In this study, the different sizes of bubble, the diffusion coefficient and the gas flow rate were numerically investigated by the CFD simulation. Different sizes of bubbles have minor effects on the mixing time. The diffusion coefficient has no significant effect on the mixing time but has significant influence on the characteristics of the plume regions and the velocity flow field. This study showed the importance of gas flow rate in the gas stirred ladle with dual plugs. It was found that gas flow rate has significant effects on turbulent kinetic energy, the mixing time and the steel cleanliness. Therefore, it is important for the steel plant to optimize the gas flow rate in order to achieve both productivity and quality of the liquid steel product.

5. References

[1] H. Liu, Z. Qi and M. Xu 2011 Numerical Simulation of Fluid Flow and Interfacial Behavior in Three-phase Argon-Stirred Ladles with One Plug and Dual Plugs, *steel research int.* **82** No. 4 pp. 440-458

[2] D. Mazumdar and J. W. Evans 2009 *Modeling of Steelmaking Processes*, Taylor & Francis Group, 451 p.

[3] H. M. Issa 2016 Power Consumption, Mixing Time, and Oxygen Mass Transfer in a Gas-Liquid Contactor Stirred with a Dual Impeller for Different Spacing, *Hindawi Publishing Corporation Journal of Engineering* **2016** pp.1-7

[4] Yan. Liu, Zhu He, and Li. Pan 2014 Numerical Investigations on the Slag Eye in Steel Ladles, *Hindawi Publishing Corporation Advances in Mechanical Engineering* **2014** pp.1-6

[5] D. Mazumdar and J. W. Evans 2004 Macroscopic Models for Gas Stirred Ladles, *ISIJ International* **44** No. 3 pp. 447-461

[6] D. Mazumdar and R. I. L. Guthrie 1995 The Physical and Mathematical Modelling of Gas Stirred Ladle Systems, *ISIJ International* **35** No. 1 pp. 1-20

[7] D. Mazumdar, H.B. Kim & R.I.L. Guthrie 2000 Modelling criteria for flow simulation in gas stirred ladles: experimental study, *Ironmaking & Steelmaking* **27** Issue 4 pp. 302-309

[8] D. Mazumdar, P. Dhandapani and R. Sarvanakumar 2017 Modeling and Optimisation of Gas Stirred Ladle Systems, *ISIJ International* **57** No. 2 pp. 286-295

[9] M. Madan, D. Satish and D. Mazumdar 2005 Modeling of Mixing in Ladles Fitted with Dual Plugs, *ISIJ International* **45** No. 5 pp. 677-685

Acknowledgments

The authors would like to thank the fund support from King Mongkut's University of Technology North Bangkok, contract no. KMUTNB-61-DRIVE-014, Millcon Steel PLC, the software license support from Flow Science, Inc and Design Through Acceleration Co., Ltd.

IOP Conf. Series: Materials Science and Engineering **526** (2019) 012029 doi:10.1088/1757-899X/526/1/012029

Synthesis of Na-A zeolite from water treatment residue using hydrothermal and microwave techniques

B Theptham[1] and S Asavapisit[1,*]

[1] Division of Environmental Technology, School of Energy, Environment and Materials, King Mongkut's University of Technology Thonburi, Bangkok 10140, Thailand.

*Corresponding author: suwimol.asa@kmutt.ac.th

Abstract. Na-A zeolite was synthesized from water treatment residue (WTR) using hydrothermal and microwave techniques. WTR was brought from Metropolitan Waterworks Authority (Thailand) and was activated by firing at 700 °C for 3 h in an electric furnace. The activated WTR was reacted with 3 M NaOH solution in a reactor under a controlled temperature of 80, 90 and 100 °C for 8, 10, 12, and 14 h. SEM (Scanning electron microscopy) images reveal the cubic crystal of zeolite. The highest cation exchange capacity (CEC) was obtained from sample synthesized at 90 °C for 14 h and gave the CEC value of 3.14 meq/g Ca^{2+}. For the microwave technique, the raw materials were reacted at 80 90 and 100°C for 30, 60, and 120 min. Results showed that the sample obtained at 80 °C for 30 min represented the highest CEC value of 0.29 meq/g Ca^{2+}.

1. Introduction

The waste from water treatment is a water treatment residue by coagulants and other chemical reagents in the purification process. The number of residue from the process about 100 ton/day and removal of WTR has carried out lead drilling for as much as 11 million baht a year [1]. The Metropolitan Waterworks Authority (Thailand) has studied to develop WTR in agriculture and industry. The main chemical components of WTR are silica, alumina, ferric oxide and lime. The substances of zeolite are almost of silica and alumina and these residues are high of silica and alumina ratio of 1-2 [2] the raw material is appropriate for zeolite synthesis. The synthesis of zeolite Na-A (Na_{12} $[(AlO_2)_{12}(SiO_2)_{12}].27H_2O$) is a porous aluminosilicate compound with numerous excellent properties, such as non-toxicity, high porosity, good thermal stability, high Ca^{2+} exchange capacity, and functional molecular sieve [3] which is generally synthesized as hydrothermal synthesis can be effected under temperatures and pressures below the critical point [4] and Microwave technique has a short time to crystallization and homogeneous nucleation [5]. The objective of the study was to compare the properties of Na-A zeolite by the hydrothermal technique with microwave heating.

2. Material and methods

2.1. Raw material and chemicals

The raw WTR samples were collected from the Metropolitan Waterworks Authority (Thailand). The composed of WTR Si/Al molar ratio is around 1.78. Sodium Hydroxide (3M NaOH) as alkali hydroxide is the used material for the preparation of Na-A zeolite. The solution for cation exchange capacity (CEC) tests using complexometric titration with EDTA.

2.2. Experimental

2.2.1. Synthesis of Na-A zeolite. WTR was crushed into different particle size and calcination at 700 °C for 3 h. The sample with the particle size larger than 325 mesh sieve hole was used to synthesize. For hydrothermal synthesis WTR obtained undergo alkaline treatment by mixing 160 g of WTR with 800 ml of 3M NaOH at 80, 90, and 100 °C for 8, 10, 12, and 14 h under stirring at 500 rpm. In the microwave synthesis at 1000 W the raw materials were reacted at 80 90 and 100°C for 30, 60, and 120 min. At the end of synthesis, the reaction mixtures were filtered and washed with distilled water until the pH of the filtrate is below 9. Then the samples were oven dried overnight at 105 °C.

2.2.2. Sample characteristics. The Na-A were characterized using different analytical techniques. The morphology of Na-A was studied using Scanning Electron Microscopy (SEM). The chemical characterization was performed by the X-ray fluorescence (XRF) and Cation exchange capacity (CEC).

3. Results and Discussion

3.1 The chemical characterization of WTR

The chemical compositions of the WTR analysis data from XRF show that in Table 1. The mainly composed of WTR is SiO_2 54.200 wt% and Al_2O_3 30.400 wt%. The Si/Al molar ratio is around 1.78. The components of WTR can be used in the synthesis of zeolites, which is quite close to the Si/Al ratio 1-2 in Na-A [2]

Table1. Chemical compositions of WTR powder

Ingredients	Composition (wt.%)
SiO_2	54.200
Al_2O_3	30.400
Fe_2O_3	8.500
K_2O	2.300
MgO	1.030
TiO_2	0.885
CaO	0.759
SO_3	0.538
P_2O_5	0.527
Na_2O	0.254
LOI	0.33

3.2 Characterization of zeolite Na-A

3.2.1. Cation exchange capacity. The cation exchange capacity of all synthetic products was evaluated and expressed in $CaCO_3$ (meq/g) anhydrous zeolite Na-A from the solution having the hardness of $CaCO_3$. The highest cation exchange capacity in the hydrothermal process shows that in Table 2 is about 3.14 meq/g obtained in To explore the effect of crystallization on the temperature rose from 80 to 100 ^oC zeolite Na-A could not be obtained until the temperature was below than 90 ^oC. When the temperature was above 90 ^oC CEC value are decreasing steadily [6].

Table2. Cation exchange capacity of Na-A zeolite by hydrothermal process

Time (h)	CEC (meq/g)		
	80°C	90°C	100°C
8	0.28	1.46	1.38
10	0.30	2.47	1.54
12	0.37	2.97	0.99
14	0.38	3.14	0.60

The sample synthesized from WTR at 90 ^oC for 14 h. The synthetic products obtained by microwave exhibited in Table 3 the maximum CEC at 80 ^oC for 30 min. When the crystallization time was the longer reaction the poorly developed Na-A would be obtained, thus the crystallization time of 30 min should be maintained to improve the crystallinity and avoid phase transformation [7].

Table3. Cation exchange capacity of Na-A zeolite by the microwave process

Time (min)	CEC (meq/g)		
	80°C	90°C	100°C
30	0.29	0.13	0.13
60	0.19	0.12	0.12
120	0.12	0.11	0.10

3.2.2. Scanning electron microscope. The size and shape of crystals of minerals can be verified by SEM micrographs of the raw materials and the end product obtained from the zeolite process. The SEM micrographs of the WTR reveal the presence of among of amorphous in Fig. 1A. The morphology and particle size of the optimum samples are shown in Fig. 1B and 1C. The SEM micrographs for the best hydrothermal synthesis zeolite Na-A at 90 ^oC for 14 h showed the formation of zeolite Na-A crystal of cubic shape (Fig. 1B). From the micrographs of the best microwave synthesis at 80 ^oC for 30 min it is possible to see the presence of zeolite Na-A crystal of cubic shape with a rounded edge. Thus, the synthesis time of crystallization by microwave synthesis was shorter than hydrothermal synthesis. Moreover, zeolite Na-A synthesized by hydrothermal is composed of well-shaped cubic crystals, while Na-A under pulsing microwave consists of sphere grains without well- developed crystal faces [8].

Figure 1. SEM micrographs of (A) WTR, (B) synthesis Na-A prepared by hydrothermal, (C) microwave.

4. Conclusions

The synthesis of zeolite Na-A from water treatment residue was done by the hydrothermal and microwave technique of synthesis. The hydrothermal synthesis gave cubic crystal of zeolite Na-A having optimum CEC at 90 °C for 14 h. Whereas, the microwave synthesis yielded cubic crystal of zeolite with cubic shape with rounded edge and CEC high-value capacity at 80 °C for 30 min. Hence, the crystallization time by microwave was shorter than hydrothermal synthesis, while zeolite Na-A synthesized by hydrothermal is composed of well-shaped cubic crystals. Therefore, hydrothermal synthesis also has the advantage of the operating alkaline process variables to achieve higher yield, purity, cation exchange capacity and crystallinity. The hydrothermal method can obtain the high temperature and high pressure in a closed reaction systems, it takes a long time to complete the reaction. Nevertheless, the microwave irritation as energy sources could lead to homogenous nucleation and uniform distribution of heat leading to a decrease in the reaction time from days down to hours and even minutes. The disadvantages of microwave irradiation are from its limitations and preparation.

5. References

[1] Metropolitan Waterworks Authority 2015, *Annual Report 2015* Metropolitan Waterworks Authority. Thailand.
[2] Cubillas P and Anderson M W 2009 Synthesis Mechanism: Crystal Growth and Nucleation *Zeolites and Catalysis* vol 1, ed Čejka J, Corma A and Zones S (WILEY-VCH Verlag GmbH & Co. KGaA) p 23
[3] Xu R, Pang W, Yu J, Hao Q and Chen J 2007 Cation-exchange Method *Chemistry of Zeolites and Related Porous Materials*, (John Wiley & Sons) p 380
[4] Morris R E 2009 Hydrothermal, Solvothermal, and Ionothermal Synthesis *Zeolites and Catalysis* vol 1, ed Čejka J, Corma A and Zones S (WILEY-VCH Verlag GmbH & Co. KGaA) p 89
[5] Xu R, Pang W, Yu J, Hao Q and Chen J 2007 Crystallization of Zeolites under Microwave Irradiation *Chemistry of Zeolites and Related Porous Materials*, (John Wiley & Sons) p 158-159
[6] Qian T and Li J 2015 Synthesis of Na-A zeolite from coal gangue with the in-situ crystallization technique, *Advanced Powder Technology*, **vol. 26**, pp 98-104
[7] Li Y and Yang W 2008 Microwave synthesis of zeolite membranes, Microwave synthesis of zeolite membranes: A review, *J. Membrane Science*, **vol.316**, pp 6
[8] Ayele L, Pariente J P, Chebude Y and Díaz I 2016 Conventional versus alkali fusion synthesis of zeolite A from low grade kaolin, *Applied Clay Science*, **vol. 132-133**, pp 485-490

Acknowledgments

This project was supported by the School of Energy, Environment and Materials and Faculty of Science, King Mongkut's University of Technology Thonburi.

IOP Conf. Series: Materials Science and Engineering **526** (2019) 012030 doi:10.1088/1757-899X/526/1/012030

Thermal and Melting track Simulations of Laser Powder Bed Fusion (L-PBF)

P Ninpetch[1], P Kowitwarangkul[1,*], S Mahathanabodee[2], R Tongsri[3], and P Ratanadecho[4]

[1] The Sirindhorn International Thai-German Graduate School of Engineering (TGGS), King Mongkut's University of Technology North Bangkok (KMUTNB), Bangkok, 10800, Thailand
[2] Department Production Engineering, Faculty of Engineering, King Mongkut's University of Technology North Bangkok (KMUTNB), Bangkok, 10800, Thailand
[3] Particulate Materials Processing Technology Laboratory (PMPT), National Metal and Materials Technology Center, Pathum Thani 12120, Thailand
[4] Department of Mechanical Engineering, the Faculty of Engineering, Thammasat University (Rangsit Campus), Pathum Thani 12120, Thailand

* Corresponding author: pruet.k@tggs.kmutnb.ac.th

Abstract Laser powder bed fusion (L-PBF) process involves with the construction of phase transformation, melting, and rapid solidification of weld metal powder bed which affects the properties and the microstructure of final parts, e.g. density, dimension, mechanical properties, void, porosity, and non-fully melted particle. The aims of this work were to study the effect of process parameters, e.g. laser power and scanning speed, on the temperature field and melt pool geometry and the characteristics of single melting track in the L-PBF process by using the commercial CFD software simulation Flow-3D (Flow-weld). The laser power, scanning speed, laser spot diameter, and layer thickness varied in this study were 120 W, 140 W, 0.4 m/s, 0.6 m/s, 0.8 m/s, 80 µm and 50 µm respectively. The results stated that at the lower scanning speed, the temperature field has a region of heat distribution larger than that of the higher one. The geometry of melt pools can be changed from ellipse shape to tear drop shape when the scanning speed is increased. The width and depth of laser melting track is increased when the higher laser power and lower scanning speed are applied. The void is found underneath the laser melting track when the scanning speed changes from 0.4 m/s to 0.6 m/s.

1. Introduction

Laser Powder Bed Fusion (L-PBF) is a type of additive manufacturing process which uses laser source scanning as a moving heat source on powder bed to fully melt the metal powder layer fashion [1]. In this process, the powder bed is created by raking metal powder across the work space [2]. The cycle is repeated layer by layer, until the complete part is formed. Finally, the part is visible after eliminated the excess metal powder from powder cake [3]. This process can be used with various materials such as metal, plastic, ceramic, composites, and biomaterials [2]. The typical applications of this process are rapid prototype, complex topology parts, medical parts, aerospace parts and automotive parts, [4,5] The process parameters include laser power, power intensity, scanning speed, hatch spacing and the layer thickness. L-PBF process involves complex multiphysics such as the materials laser absorption, heat transfer, fluid dynamic, phase transformation, melting, and rapid solidification. These factors have significant influence on the properties and the microstructure of final parts, e.g. density, dimension, mechanical properties, void, porosity, and non-fully melted particle [6]. Several research studies used the CFD simulation method to predict the phenomena in the L-PBF process. The examples of CFD simulation software are Flow-3D, ANSYS, COMSOL Multiphysics etc. Y. Li and D. Gu [7] developed a 3D model to study thermal behavior of L-PBF of pure titanium powder with ANSYS software. The study considered the temperature-dependent thermos physical parameters and the Gaussian distribution of the heat flux. In the study, it indicated that the average temperature of the powder bed gradually increased during laser scanning because of the heat accumulation. Y.S. Lee and W. Zhang [6] solved the transient numerical mesoscale model of Inconel 718 powder bed with a mesh size of 3 micron considering volume of fluid method (VOF), Marangoni effect, particle size distribution and packing density. The commercial software Flow-3D was used in the study. It was found that too fast scanning speed and too low laser power can increase the chance of discontinuous molten pool and the balling defect.

The aims of this research are to study the effect of process parameters including Laser power and scanning speed on the temperature field and melt pool geometry and the characteristics of single melting track in the L-PBF process by using the commercial CFD software simulation Flow-3D (Flow-weld). The laser power, scanning speed, laser spot diameter, and layer thickness varied in this study are 120 W, 140 W, 0.4 m/s, 0.6 m/s, 0.8 m/s, 80 μm and 50 μm respectively. The results from this study will be validated with the experimental results from a previous research study on Fabrication and Characterization of AISI 420 Stainless Steel Using Selective Laser Melting [8].

2. Simulation model

Heat transfer simulation model in L-PBF process consists of heat conduction, heat convection, heat radiation respectively [9]. Considering the melting and solidification phenomena, the enthalpy including latent heat evolution arising from phase change effect during the process can be applied in modelling [10]. The molten fluid flow simulation model of L-PBF process based on numerical solution of mass, energy and momentum conservation and the Volume of Fluid (VOF) method is used to track the position and shape of the molten pool surface.

The Gaussian laser moving heat source model is described [11] as in equation (1)

$$q = \frac{2P}{\pi\omega^2} e^{-\frac{[(x-x_0)^2+(y-y_0)^2+(z-z_0)^2]}{\omega^2}} \tag{1}$$

q is heat flux on the surface, ω is radius of beam, P is laser power, (x_o, y_o, z_o) is instantaneous position of the center of the heat flux which is on the moving path at distance of x, y, z from start point, v is velocity of moving heat source, t is time.

3. Methodology

The 3D model simulation was performed by commercial CFD simulation software Flow-3D (Flow-weld). The material used in this study was AISI 420 martensitic stainless steel powder with the average particle size of 20 μm. The 3D computational mesh of around 375,000 cells was used in this simulation. Figure 1. shows the schematic of the powder bed and scanning direction of moving laser beam. The process parameters used in this study are listed in Table 1.

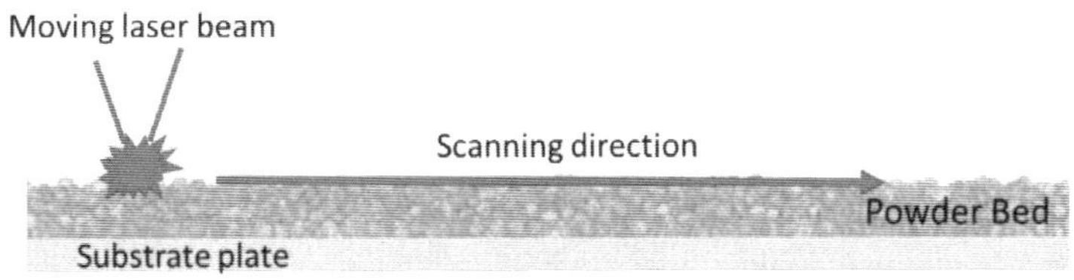

Figure 1. The schematic of the powder bed and scanning direction of moving laser beam

Table 1. Process parameters of this simulation

Process parameters	Value
Laser power (P)	120, 140 Watt
Scanning speed (v)	0.4, 0.6, 0.8 m/s
Layer thickness (t)	50 μm
Laser beam spot size (d)	80 μm

4. Results and Discussion

4.1 The effects of process parameters on the temperature field and melt pool shape

At the lower scanning speed, the temperature field has a larger region of heat distribution than that of the higher one as shown in figure 2 **(a), (b).**When the scanning speed is increased, the geometry of melt pools changes from ellipse shape to tear drop shape**.** The phenomena from the process parameters have significant effect on the quality of final products**.**

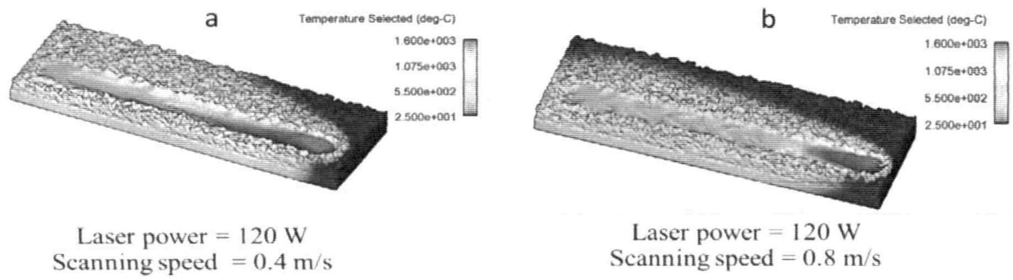

Figure 2. The temperature field and melt pools shape during L-PBF process

4.2 The width and depth of single melting track with different process parameters

Figure 3 (a), (b) and figure 4 (a), (b) show the effects of laser power and scanning speed on the characteristics of laser melting track**.** The width and depth of laser melting track is increased when the higher laser power and lower scanning speed are applied**.** The void is found underneath the laser melting track when the scanning speed changes from 0.4 m/s to 0.6 m/s**.** The results from the simulation are consistent with the experimental results with minor difference as shown in figure 4**.** (a)

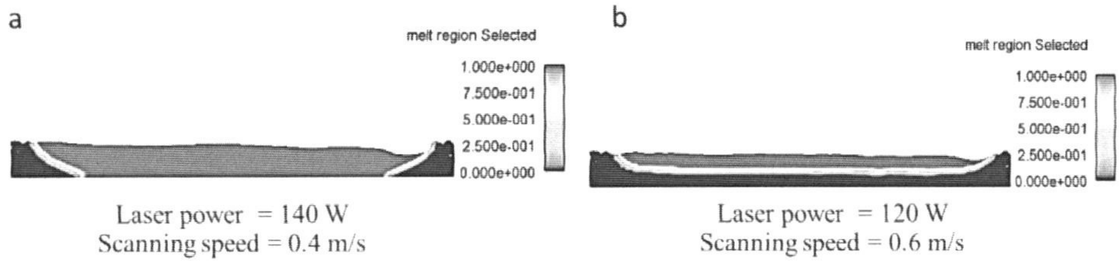

Figure 3. The depth of laser melting track in L-PBF process

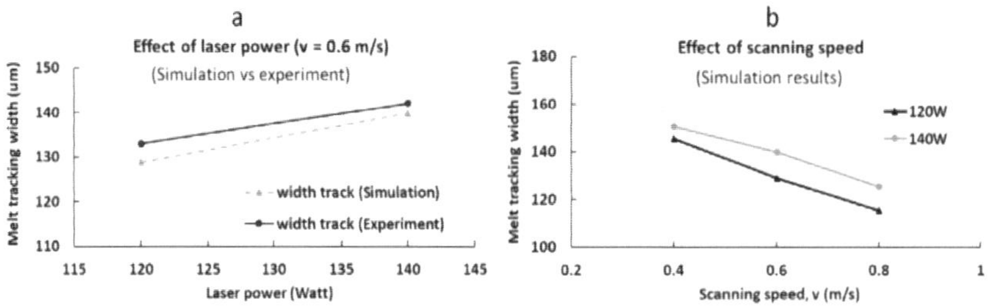

Figure 4. The effect of process parameters on melting track width; (a) The effect of laser power (b) the effect of scanning speed

5. Conclusions

In this research, the effect of process parameters on the temperature field and the melt pool geometry and the characteristic of single melting track in L-PBF process with laser power of around 120-140W was investigated through the numerical simulation method. The results from the simulation are consistent with the experimental results with minor difference. At the lower scanning speed, the temperature field has a region of heat distribution larger than that of the higher one. The geometry of melt pools can be changed from ellipse shape to tear drop shape when the scanning speed is increased. The width and depth of laser melting track is increased when the higher laser power and lower scanning speed are applied. The void is found underneath the laser melting track when the scanning speed changes from 0.4 m/s to 0.6 m/s.

6. References

[1] ASTM Standard F-2792- 12a. Standard Terminology for Additive Manufacturing Technologies, ASTM International

[2] W. E. Frazier 2014 Metal Additive Manufacturing: A Review, *Journal of Materials Engineering and Performance* Vol **23**, pp. 1917–1928

[3] V. Bhavar, P. Kattire, V. Patil, S. Khot, K. Gujar and R. Singh 2014 A Review on Powder Bed Fusion Technology of Metal Additive Manufacturing, *International conference and exhibition on Additive Manufacturing Technologies*-AM-2014, September 1 &2 ,2014, Banglore, India.

[4] T. Wohlers 2010 Additive Manufacturing State of the Industry, Wohlers report, pp. 1-26

[5] C. H. Fu and Y. B. Guo, 3-dimension finite element modelling of selective laser melting Ti-6AL-4V alloy, *25th Annual International Solid Freeform*, pp. 1129-1144

[6] Y. S. Lee and W. Zhang 2015 Mesoscopic Simulation of Heat Transfer and Fluid Flow in Laser Powder Bed Additive Manufacturing, *International Solid Free Form Fabrication Symposium Austin*, pp. 1154-1165

[7] Y. Li and D. Gu 2014 Thermal behavior during selective laser melting of commercially pure titanium powder: Numerical simulation and experimental study, *Additive Manufacturing* Vol **1–4**, pp. 99–109

[8] X. Zhao, Q. Wei, B. Song, Y. Liu, X. Luo, S. Wen, and Y. Shi 2015 Fabrication and Characterization of AISI 420 Stainless Steel Using Selective Laser Melting, *Materials and Manufacturing Processes*, pp. 1–7

[9] K. Zeng, D. Pal and B. Stucker 2012 A review of thermal analysis methods in Laser Sintering and Selective Laser Melting, *Journal of Heat Transfer*, pp. 796-814

[10] A. M. Kamara, W. Wang, S. Marimuthu, and L. Li 2010 Modelling of the melt pool geometry in the laser deposition of nickel alloys using the anisotropic enhanced thermal conductivity approach, *Proc. IMechE* Vol. **225**, pp. 87-99

[11] A. Srivastava, Moving heat Source Version 4.1, Technical document, https://appstore.ansys.com, ANSYS, Inc, accessed on 1 Aug 2017

Acknowledgement

The authors would like to thank the fund support from Thailand Graduate Institute of Science and Technology (TGIST), National Science and Technology Development Agency (NSTDA) contract No. SCA-CO-2560-4552-TH, the Thailand Research Fund (TRF) and the Commission on Higher Education (CHE) under the TRF contract No. MRG6180164 and software license support from Flow Science, Inc, Flow Science Japan and Design Through Acceleration Co., Ltd.

Vinyl *Cis* Rubber superior polymer for future trends

T Puvanatvattana[1,*], and T Chansorn[1]

[1] UBE Technical Center (Asia) Limited, Rayong, Thailand, 21000

*Corresponding author: toemphong@ube.co.th

Abstract. Vinyl *Cis* Rubber (VCR) is *cis*-1,4-polybutadiene reinforced by highly crystalline fibrous syndiotactic-1,2-polybutadiene (SPB). It has been developed and commercialized by Ube Industries Co.,Ltd. The highly crystalline syndiotactic-1,2-polybutadiene having melting point (M.p.) up to 473K is obtained by using cobalt catalyst system. Fibrous crystalline domains in VCR are effective for the stress-induced crystallization of the rubber matrix. VCR has various improved properties such as high green strength, good processing, large modulus, good flex-crack resistance, etc. The improvement of physical and dynamic mechanical properties depends on the content of syndiotactic-1,2-polybutadiene dispersion in *cis*-1,4-polybutadiene matrix and rubber compound recipe. For examples; with the same compound recipe, VCR vulcanized significantly improve in modulus up to 80%, increase hardness up to 10%, and improve in flex-crack resistance. With the same hardness of vulcanized recipe, filler can be reduced down to 10-20 part per hundred rubbers consequently energy loss performance can be improved up to 10-18%. This polymer is widely used in eco-friendly tires, high performance tires, out sole of shoe, conveyor belts and other rubber goods.

1. Introduction

Regarding to tire labeling regulation have been started since 2012 that must specify fuel consumption, wet grip and noise classification of every tire sold in market. This regulation designed to improve the energy performance of tires, buildings and services to reduce energy consumption by 20% until 2020 [1].

Polybutadiene rubber is one important component of tire both of passenger car tires and truck bus tires. The design of polymer structure of polybutadiene, for example, linearity, molecular weight or molecular weight distribution is importance for producing suitable rubber with lowers energy loss and higher abrasion resistance. Furthermore, the composite polybutadiene rubber is one of technology that can provide satisfying properties of rubber composition for eco-friendly and high-performance tires. Such a composite polybutadiene is commercially available under registered trade mark UBEPOL VCR® from composite technology of Ube Industries Co., Ltd. This polymer in the form of a composite composed of an elastomer which is a part of *cis*-1,4-polybutadiene and fibrous crystals which is a part of syndiotactic-1,2- polybutadiene uniformly dispersed in elastomer matrix with desired amount [2]. UBEPOL VCR412, its composition includes *cis*-1,4-polybutadiene 88% and syndiotactic-1,2-polybutadiene 12%. UBEPOL VCR617, its composition includes *cis*-1,4-polybutadiene 83% and syndiotactic-1,2-polybutadiene 17%, respectively.

The present study, the suitable rubber compositions comprises a butadiene rubber containing syndiotactic crystal will be provided for using in standard typical sidewall formulation of pneumatic tires, which is having not only high reinforcement, high durability, and excellent in processability but also having satisfy energy loss performance.

2. Experimental

In this study, two cases were investigated; same formulation and same hardness of vulcanizate formulation.

2.1. Materials

2.1.1 Natural Rubber

2.1.2 Polybutadiene rubber. Polybutadiene rubber containing syndiotactic crystal; UBEPOL VCR412, UBEPOL VCR617, Neodymium polybutadiene rubber (NdBR)
2.1.3 High Abrasion Furnace carbon black (HAF, N330), other standard rubber compounding ingredients such as stearic acid, zinc oxide, processing oil, accelerator and sulfur are commercial grades.

2.2 Mixing and curing

To prepare the compounds, rubber compounds were mixed in small banbury mixer with different compound formulation as shown in Table 1. Preparation of sheets for testing was done on a two roll mill.

Ingredients (phr)	S-1	S-2	S-3	S-4	S-5	S-6
UBEPOL VCR412	50					
UBEPOL VCR412		50				
UBEPOL VCR617			50			
UBEPOL VCR617				50		
NdBR-1					50	
NdBR-2						50
STR5L	50	50	50	50	50	50
CB N330	30	50	30	50	50	50
Process oil	10	10	10	10	10	10
ZnO	3	3	3	3	3	3
Stearic acid	2	2	2	2	2	2
Antioxidant	2	2	2	2	2	2
TBBS	0.8	0.8	0.8	0.8	0.8	0.8
Sulfur	1.5	1.5	1.5	1.5	1.5	1.5
Total	149.3	189.3	149.3	189.3	169.3	169.3

Table 1 Rubber compound formulation

For each type of rubber compounds, the vulcanization process was performed by compression molding process. The optimal curing time (t90) obtained from RPA2000 Rheometer (Alpha Technologies) measurements according to ASTM D5289

3. Results and discussion

3.1 Rubber compound properties

It was found that the Mooney viscosity of rubber compounds containing same amount of carbon black were increased related with the Mooney viscosity of polybutadiene rubber as shown in Table 2. Both of rubber compounds containing SPB showed lower Mooney viscosity than NdBR. The results were inherently related to the processability test which rubber compounds included UBEPOL VCR showed better workability, smoother surface of extrudate, and smaller die swell than NdBR as shown in Figure 1. This phenomenon is corresponding to reinforcement effect of syndiotactic crystal in rubber compounds.

The carbon black content was adjusted in order to keep same hardness level of vulcanizates at 60 Shore A in this study. It was found that Mooney viscosity of rubber compound containing SPB were spontaneously decreased with decreasing amount of carbon black as shown in Table 2. Lower filler contents showed lower compound viscosity that inherently related to better workability. Rubber compounds containing UBEPOL VCR still showed better processability than NdBR compounds as shown in Figure 1. This phenomenon is corresponding to reinforcement effect of syndiotactic crystal in rubber compounds as discussion above.

Table 2 Raw polymer properties of each polybutadiene rubber

Property	UBEPOL VCR412	UBEPOL VCR412	UBEPOL VCR617	UBEPOL VCR617	NdBR-1	NdBR-2
Carbon Black (phr)	30	50	30	50	50	50
Raw Polymer ML1+4 (373K)	44	44	64	64	66	49
Syndiotactic-1,2-polybutadiene (%)	12	12	17	17	-	-
Rubber compound ML1+4 (373K)	40	56	43	62	69	61
Cure characteristic, 423K t10 (min)	8.1	7.0	8.1	6.3	6.9	7.0
t90 (min)	15.0	13.5	14.7	12.9	13.4	13.6
Extrusion Die Swell	2.61	2.31	2.20	2.14	3.00	2.94

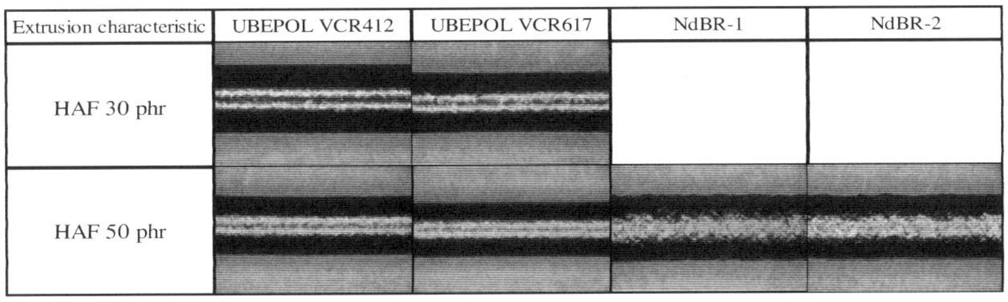

Figure 1 Extrusion characteristic of each rubber compound at shear rate 2,000 (1/sec)

3.2 Vulcanizate properties

We observed that rubber compounds containing SPB showed higher modulus than NdBR compounds around 19%-85% as shown in Table 3. This behavior is related to reinforcement effect of syndiotactic crystal in rubber compound. The effect is more evident in the case of higher syndiotactic crystal content. Furthermore, syndiotactic crystal containing compounds showed better Picro abrasion resistance and crack growth resistance than NdBR compounds. This behavior can be explained that syndiotactic crystal act as fibrous filler well dispersed and interact with polymer, consequence of higher hardness and Picro abrasion resistance. In addition, syndiotactic crystal can be also seen as rigid polymer-filler interface which leads to accumulation of stress and crack can change direction of growth, which may lead to crack stopping [3].

It was also observed that carbon black content could be reduced 10-20 phr for rubber compounds containing SPB in order to keep the same hardness level as NdBR compounds. However, they showed same modulus level as NdBR compounds even lower carbon black content, especially UBEPOL VCR617. This behavior is related to reinforcement effect of syndiotactic crystal as mention above. The decreasing filler content leads to increase the mobility of rubber chains, consequently, induce elasticity in lower filled compounds [4]. Therefore, elasticity of rubber compounds containing SPB was improved. Their rebound resilience was same level as NdBR compounds. In addition, heat generations of low filled UBEPOL VCR compounds were better than NdBR compounds. Furthermore, these compounds still kept better performance of crack growth resistance than NdBR rubber compounds, especially, in low stoke area. These behaviors of syndiotactic crystal containing compounds will be suitable for sidewall application of future eco-friendly and high performance tires.

The dynamic viscoelastic properties for the low filler filled rubber containing syndiotactic crystal compounds are shown in Table 3. It was observed that syndiotactic crystal containing compounds with lower filler content still showed higher tanδ in low temperature corresponds to better wet skid resistance compared to NdBR compounds. In addition, their rolling resistance performance which related to tanδ in the temperature range of 333K was better than NdBR compounds. It can be explained that rubber compounds containing SPB will dissipate and lose more energy in the low temperatures range. Therefore, these rubber compounds have better wet skid resistance. However, filler loading decreases the filler-filler interaction decreases, consequently, reduce energy loss from filler-filler friction in the compounds.

Table 3 Mechanical properties of each vulcanizates

Testing items	UBEPOL VCR412	UBEPOL VCR412	UBEPOL VCR617	UBEPOL VCR617	NdBR-1	NdBR-2
Mechanical properties						
Carbon black (phr)	30	50	30	50	50	50
Specific Gravity (g/cm^3)	1.046	1.099	1.047	1.100	1.098	1.097
Hardness (Shore A)	55	63	61	68	60	60
100% Modulus (MPa)	1.7	2.5	2.5	3.9	2.1	2.1
200% Modulus (MPa)	4.0	7.0	5.5	9.5	5.6	5.5
300% Modulus (MPa)	7.5	12.6	9.4	15.3	11.1	10.7
Tensile at Break (MPa)	22.1	21.9	19.4	22.3	22.9	23.9
Elongation at Break (MPa)	608	494	516	462	515	545
Tear Resistance (kN/m)	65	73	68	72	72	71
Rebound (%)	65	58	64	57	67	65
Compression set (%)	28.8	29.8	31.7	40.1	27.4	28.8
Heat Build Up : ΔT (K)	13.5	21.8	15.0	24.5	16.8	18.0
Picro Abrasion (Index)	97	152	102	167	140	142
Crack growth resistance (mm/cycles)						
Stroke 30 mm ($\times 10^{-6}$)	0.61	1.20	1.10	1.24	4.47	5.41
Stroke 56 mm ($\times 10^{-4}$)	3.59	4.15	3.43	0.10	2.94	1.07
Viscoelastic properties						
Carbon black (phr)	30	50	30	50	50	50
tanδ @ 253 K	0.237	0.267	0.241	0.260	0.214	0.222
tanδ @ 273 K	0.132	0.173	0.142	0.185	0.125	0.139
tanδ @ 333 K	0.095	0.148	0.105	0.131	0.109	0.115

4. Conclusion

UBEPOL VCR is polymer composite include mixture of *cis*-1,4-polybutadiene and a short fiber of crystalline syndiotactic-1,2-polybutadiene. Syndiotactic crystal containing compounds showed better workability, higher modulus up to 80%, higher hardness, better Picro abrasion resistance, better crack growth resistance and better wet skid resistance performance than non-containing rubber compounds in same compound formulation. In order to keep the same hardness level of vulcanizates, filler contents could be reduced 10 to 20 phr in syndiotactic crystal containing compounds formulation. Low filled rubber compound of syndiotactic crystal showed better workability, better heat generation, better wet skid resistance performance and better rolling resistance performance up to 10% to 18% than non-containing rubber compounds. These behaviors of syndiotactic crystal containing compounds are useful for sidewall application of eco-friendly and high performance tires as future trends.

5. References

[1.] http://en.wikipedia.org/wiki/Tire_label, December 17, 2018.

[2.] Muraoka et al. Rubber composition for reinforcing bead of tires which contains butadiene rubber with syndiotactic crystals, carbon black, sulfur and sulfonamide accelerator. US005859142A, Sumitomo Rubber Industries, Ltd.

[3.] E. Cichomski (Sp), W. K. Dierkes et al. Influence of Physical and Chemical Polymer-filler Bondson Wet Skid Resistance and Related Properties of Passenger Car Tire Treads, July, 2012

[4.] A. Mostafa, A. Abouel-Kasem, M.R. Bayoumi, M.G. El-Sebaie. Effect of carbon black loading on the swelling and compression set behavior of SBR and NBR rubber compounds, Materials and Design 30 (2009) 1561–1568

[5.] Fröhlich J, Niedermeier W, Luginsland HD. The effect of filler–filler and filler–elastomer interaction on rubber reinforcement. Composites Part A 2005; 36:449–60.

IOP Conf. Series: Materials Science and Engineering **526** (2019) 012032 doi:10.1088/1757-899X/526/1/012032

Solvated lithium ions in defective Prussian blue

N Ruankaew[1, 2], N Yoshida[2], and S Phongphanphanee[1, 3, 4]

[1] Department of Material Science, Faculty of Science, Kasetsart University, Bangkok 10900, Thailand.
[2] Department of Chemistry, Graduate School of Science, Kyushu University, Fukuoka 819-0395, Japan.
[3] Computational Biomodelling Laboratory for Agricultural Science and Technology (CBLAST), Faculty of Science, Kasetsart University, Bangkok 10900, Thailand.
[4] Specialized Center of Rubber and Polymer Materials in Agriculture and Industry (RPM), Faculty of Science, Kasetsart University, Bangkok 10900, Thailand.

*Corresponding authors: noriwo@chem.kyushu-univ.jp and fscisrph@ku.ac.th

Abstract. Prussian Blue (PB) or ferric hexacyanoferrate (II) is one of the most attractive materials in green battery industry. It has defective structure (d-PB), $Fe_4^{III}[Fe^{II}(CN)_6]_3 \cdot xH_2O\ (x \approx 14-16)$, that one of $[Fe(CN)_6]^{4-}$ group is missing, and hydrated with water molecules at the 24*e* empty nitrogen site. In this study, we investigated the distributions of water and Li^+ in d-PB by using three-dimensional reference interaction site-model (3D-RISM) theory, which is a statistical mechanics theory of molecular liquids. The theory can examine the selective ion adsorption in solvated molecular materials. The results from 3D-RISM show that there are 14 water molecules located in different positions inside the d-PB lattice, which are in good agreement with experiment. Moreover, we also found the Li^+ distributed at inside d-PB lattice positioned closely to the 32*f* spherical cavity of d-PB.

1. Introduction

Since the rapid development of mobile communication device and electric automobile, the green and rechargeable battery has been of great interest to high capacity power storage. The demand on lithium-ion storage battery has increased every year, and thus significant effort has been devoted to improvement of the ion capacity, durability and the environment friendly [1]. However, many materials have been introduced and selected for targets of improvement or development of novel rechargeable ions storage. Prussian blue (PB) and its analogues (PBAs) have received attention due to great ions adsorption, rechargeable, safe, and low-cost synthesis. For these reasons, PB has been widely used in many fields of science, such as chemistry, pharmaceutics, medicine, nanomaterials, and various industrial applications [2, 3]. Recently, we have studied the basis of alkali ions adsorption and selectivity of perfect structure PB in aqueous solutions by Hartree-Fock (HF) method and the 3D-RISM [4]. Our results show the dramatically difference of the absorption sites in perfect structure PB between small and large alkali ions. However, it still does not clearly understand that the defect of PB plays role on the selective adsorption or not. In this work, we studied the solvated structure inside the cavity of d-PB by using 3D-RISM. In our calculation, the charges of the d-PB were determined by Density Functional Theory which is more suitable for metal-ligand complex than HF method [2].

2. Computational Method

A single unit cell of d-PB (figure 1a) was considered as a solute in this study. The structure was derived from the neutron scattering studied by Sharma et al. the bond distance between atoms in Fe^{2+}–C, Fe^{3+}– N, and C – N are 1.923, 2.029, and 1.131 Å, respectively [5]. We focused only on the main skeleton of d- PB by removed all those 14-16 water molecules from its crystal structure. The electrostatic potential (ESP) atomic charges in d-PB were performed by Gaussian 09 Revision B [6]. The conventional ESP fitting point procedures were employed with Beck-3LYP hybrid functional. The LanL2DZ - effective core potential basis set was applied for all Fe atoms, and 6-31G(d) basis set was rest for all C and N atoms. All calculations were employed under the single point energy without structure optimization. In the iterative SCF calculations, the option SCF = XQC was set. The parameters and condition for the solvent–solvent correlation functions in 1DRISM and calculation for solute-solvent correlation functions by 3DRISM are the same the parameters as used in reference [4].

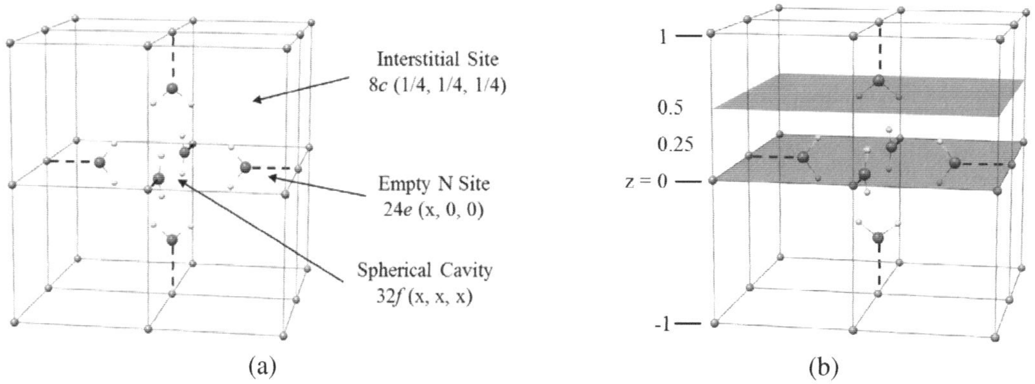

Figure 1. (a) Defective Prussian blue structure; Iron (orange), oxygen (red) and hydrogen (light blue) of water. The cyanide ligands which lie between iron atoms, the water at *32f* spherical cavity and at *8c* interstitial site were omitted for guiding your eyes. (b) The illustrated of xy contour plans (gray) on *z*-axis; z = 0, 0.25 (omitted) and 0.5.

3. Results and Discussion

3.1 Water Distribution inside d-PB

Figure 2a illustrates the isosurface plot three-dimensional distribution function (3D-DF) for oxygen atom of water (O_w) at g(**r**) > 4 inside d-PB. The term $g_{Ow}(\mathbf{r}) > 4$ in the context means that the probability of finding the O_w at position **r** is 4 times greater than that of the bulk; this term also apply to all of the others solvent species in our calculation for following discussion. The contour maps for 3D-DF of O_w at difference layer are presented in figure 2b and 2c. (The contour planes are defined as passing through the center of d-PB unit cell (*z* = 0), and through the center of subunit (*z* = 0.5), as shown in the figure 1b). The 3D-DF of O_w demonstrates the high distribution of water at the 24*e* or empty nitrogen sites, 8*c* or interstitial sites, and board peak at 32*f* spherical cavity site. Moreover, our results show the six sharp peaks of distribution of water at 24*e* (figure 2b), which indicates the strong coordination of water to Fe^{3+}. The length between the Fe^{3+} and coordinated water is about 2.13 Å, which is in good agreement with experimental study of Shores et al. [7]. In the same figure, the contour plot demonstrates the board peak at the center of the unit cell or 32*f* spherical cavity, this peak represents the non-coordinated water. Furthermore, the figure 2c shows moderately high of 8 peaks located at 8*c* interstitial sites, these peaks represent the other type of non-coordinated water in d-PB system. In other word, there are three types of water in the d-PB; (1) coordinated water (2) central cavity water and (3) interstitial water. The number of water molecule inside d-PB can be calculated by integrate the 3D-DF of water. We found that there were 14.2 water molecules distributed in d-PB, which according with the 6 coordinated water at 24*e* empty nitrogen site and 8 non-coordinated water at 8*c* and others rest inside the 32*f* spherical cavity of d-PB.

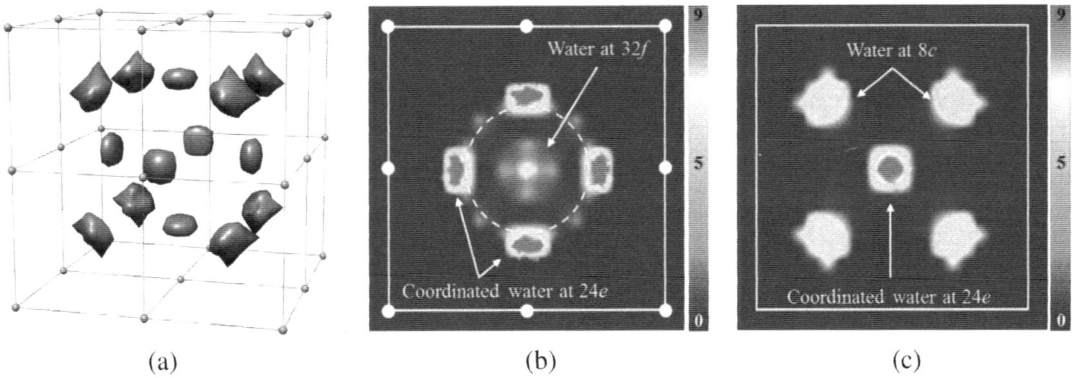

Figure 2. (a) Isosurface plot of 3D-DF of O_w at g(**r**) > 4, (b and c) the xy contour plot on plan $z = 0$ and 0.5, respectively.

3.2 Amount of water and solvated structure of explicit Li^+

The 3D-DF of Li^+ shows the highest peak located at the place in between the 8*c* interstitial site and the 32*f* spherical cavity site of d-PB; as shown in figure 3a. These peaks correspond to the adsorption sites of Li^+ in the d-PB. To determine the solvation structure of the explicit Li^+ and its coordination number, an explicit Li^+ was placed at the highest peak as shown in figure 3b and 3c, and then 3D-DF of water around the system was calculated by 3D-RISM. The results show that the explicit Li^+ is coordinated by almost four water molecules. At this position, the configuration of an explicit Li^+ and water possess a tetrahedral molecular geometry, the ion is solvated by three water molecules at 24*e* empty nitrogen site and one water molecule at 8*c* interstitial site as illustrated in figure 4a. The radial distribution function

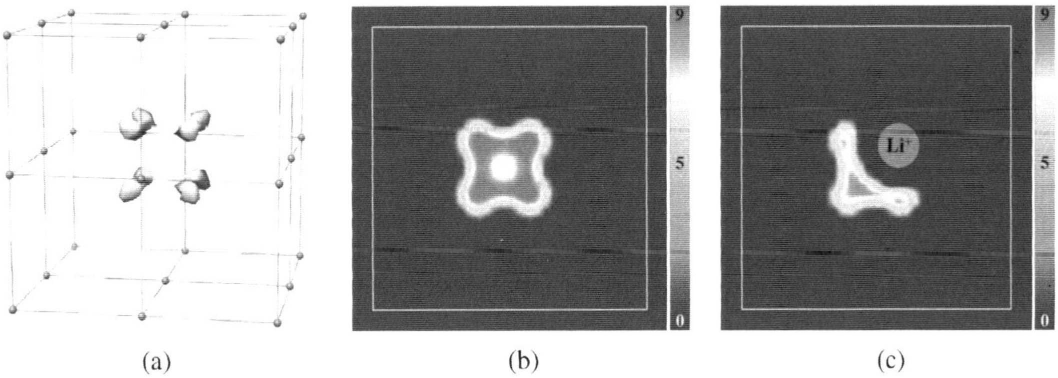

Figure 3. (a) An isosurface plot of 3D-DF of Li^+ (blue) at g(**r**) > 8 inside d-PB structure, (b) a contour plot of 3D-DF of Li^+ on plan $z = 0.25$ and (c) the contour plot on $z = 0.25$ plan after a one explicit Li^+ was placed at the most highest peak.

(RDF) of water with respect to the explicit Li^+ was calculated by angular averaging of the 3D-DF, and the coordination number of the explicit Li^+ was calculated by integration of the RDF. In figure 4b, we compare the RDF of water-Li^+ in which the ion immersed in bulk solution and the ion located at the adsorption site in d-PB. The first peak of RDF of g_{OLi} inside d-PB located at the same position with that of bulk solution. This stands for the average distance between O_w and explicit Li^+ is the same as the ions in bulk solution. Then, the total amount of water inside the system of d-PB was measured by integrating the 3D-DF of water inside the structure. The amount of water molecules in d-PB with explicit Li^+ was 14.1, when compared to the number of water molecules in the d-PB without an explicit ion was 14.2. This means that the after we placed an explicit ion the water molecule inside d-PB remains unchanged. This result indicates that there is no water lost when the Li^+ intercalated into PB; this is similar phenomenon with the case of ions adsorption of PB in our previous work [4] and in the literature [2, 8].

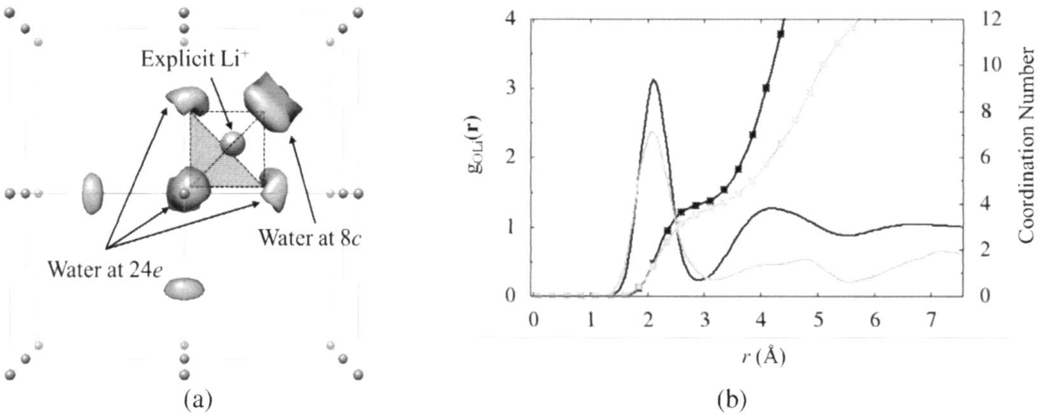

Figure 4. (a) 3D-DF of O_w (red) and explicit Li^+ (purple) which form tetrahedrally coordinated with 4 water molecules inside d-PB structure. (b) RDF between O_w and Li^+ in bulk (——) and in d-PB (——). The coordination number of Li^+ in bulk (—✱—) and in d-PB (—✱—).

4. Conclusion

We have studied the distribution of water inside d-PB and the total amount of water of the explicit Li^+ as well as its solvation structure. The results of 3D-DFs show that there are 3 types of water inside d-PB; coordinated water at 24*e* empty nitrogen site, non-coordinated water at 8*c* interstitial site and non-coordinated water at 32*f* spherical cavity site. Moreover, we found that the amount of water molecules inside d-PB were 14.2 and 14.1 molecules for the d-PB with and without adsorb one Li^+. The explicit Li^+ was coordinated by almost four water molecules by possess a tetrahedral molecular geometry.

5. References

[1] Bhatt P, Meena S S, Mukadam M D, Mandal B P, Chauhan A K, and Yusuf S M 2018 *New J Chem.* **42 (6)** 4567–4578.

[2] Hegner F S, Galán-Mascarós J R, and López N. 2016 *Inorg. Chem* **55(24)** 12851–12862.

[3] Sun X, Duffort V, and Nazar L F. 2016 *Adv Sci.* **3(8)** 1600044.

[4] Ruankaew N, Yoshida N, Watanabe Y, Nakano H, Phongphanphanee S. 2017 *Chem Phys Lett.* **684** 117–125.

[5] Sharma V K, Mitra S, Thakur N, Yusuf SM, Juranyi F, and Mukhopadhyay R. 2014 *J Appl Phys.* **116(3)** 034909.

[6] Frisch M J, Trucks G W, Schlegel H B, Scuseria G E, Robb M A, Cheeseman J R, et al. 2009 Gaussian 09, Revision B.01.Gaussian, Inc., Wallingford CT.

[7] Shores M P, Beauvais L G, Long J R. 1999 *J Am Chem Soc* **121(4)**:775–779.

[8] Paolella A, Faure C, Timoshevskii V, Marras S, Bertoni G, Guerfi A, et al. 2017 *J Mater Chem A* **5(36)** 18919–18932.

Acknowledgement

This work was supported by the Kasetsart University Research and Development Institute (KURDI).

IOP Conf. Series: Materials Science and Engineering **526** (2019) 012033 doi:10.1088/1757-899X/526/1/012033

Evaluation of silver -copper-tin brazing alloy corrosion behavior in chloride containing electrolyte at various temperature

A Nurrochman[1,2,*,§], E Prajatelistia[3,§], K Cho[4]

[1] Mining Engineering Department, Universitas Islam Bandung, Bandung, Indonesia
[2] Mechanical Engineering Department, STT Mandala, Bandung, Indonesia
[3] Materials Science and Engineering Department, Institut Teknologi Bandung, Bandung, Indonesia
[4] Materials Science and Engineering Department, Yeungnam University, Gyeongsan, South Korea

*Corresponding author: andrienurrochman@gmail.com
[§] Both authors contributed equally to this work

Abstract. Silver base brazing alloys are commonly used for brazing most of ferrous and nonferrous metals alloys. These families of filler metals have a lower brazing temperature due to Sn addition thus offers excellent flow. Meanwhile, the high content of Ag in the alloys guarantees a good corrosion resistance. However, the detrimental effect was shown when it joints in stainless steel. In present work, corrosion behavior of silver-copper-tin brazing alloy was evaluated. The brazing alloy was exposed to water contained chloride and the temperatures were varied at 25, 40, 55, and 70 °C. The electrochemical methods were utilized to evaluate the corrosion behavior of the alloy. The results showed that the dissolution of Sn seems to be initiating the corrosion of the silver-copper-tin brazing alloy. Nevertheless, the active dissolution of the Cu to the solution simultaneously influence the corrosion behavior of the silver-copper-tin brazing alloy. This study in terms of the alloy corrosion behavior in the particular environment should give both of academicals and industrials contributions.

1. Introduction

Many different types of pure metals and alloys of brazing filler materials are developed. Among them, silver base brazing alloys are commonly used for brazing most ferrous and nonferrous metals and alloys. These families of filler metals have lower brazing temperature and offer excellent flow. Moreover, the silver-copper-tin brazing alloys are free of phosphorus, the element that was believed as the catalyst of the most phosphorus-containing brazing alloy corrosion [1].

Very often failures of the braze joints occurs in the brazing alloys at first [2, 3]. The complex either electrochemical or chemical corrosion process in the brazing alloys builds an aggressive environment that harmful to the base materials. Many brazed joints consist of two or more phases. Sometimes one or more of the phases will be subjected to selective attack. Occasionally, an element is selectively leached from the braze [4]. Thus, corrosion properties of the brazing alloys must be considered when selecting the appropriate brazing alloys. Studies in terms of their behaviour in the particular environment should give both of academicals and industrial contributions.

IOP Conf. Series: Materials Science and Engineering **526** (2019) 012033 doi:10.1088/1757-899X/526/1/012033

In the present study, we intended to evaluate the corrosion behaviour of the silver base brazing alloys. This study was focused to the effect of the temperature on corrosion behaviour of silver-copper-tin brazing alloy in chloride containing electrolyte.

2. Experimental Procedures

The materials used were commercial silver-copper-tin brazing alloy and supported by Doosan Heavy Industries & Construction Ltd. The compositions of the brazing alloy are 9.5% Sn, 30.5% Cu, Ag bal. The materials were pre-treated by mechanically polished with SiC abrasion paper from 600 to 2000 grits, and then mirror finished using Buehler polishing cloth wetted with diamond suspension up to 1 µm. Afterwards, the materials were cleaned by ultrasonically degrease for 5 minutes in methanol and ethanol before rinse thoroughly in distilled water before finally dried.

The corrosion behaviour was evaluated by electrochemical method such potentiodynamic polarization, cyclic voltammetry, and potentiostatic polarization. The experiments were conducted using three-electrode cell with the reference electrode used was silver/silver chloride (Ag/AgCl saturated KCl). The condition was under open air atmosphere with varied temperature of 25, 40, 50, and 70°C in 3.5 wt% NaCl electrolyte prepared by using distilled water and analytical grade NaCl.

3. Results and discussions

The potentiodynamic polarization of the silver-copper-tin brazing alloys in 3.5 wt% NaCl solution resulted in E vs log I curves and represented in figure 1. The anodic region at 25°C showed the typical anodic polarization behaviour of Cu-Sn metal alloys in aqueous chloride media [5-8]. The curve consisted of two regions that have minimum current density. The corrosion potential (E_{corr}) of the alloy at 25°C in 3.5 wt% NaCl solution was -0.135 V_{SSCE} with the corrosion current of 3.864×10^{-7} A.

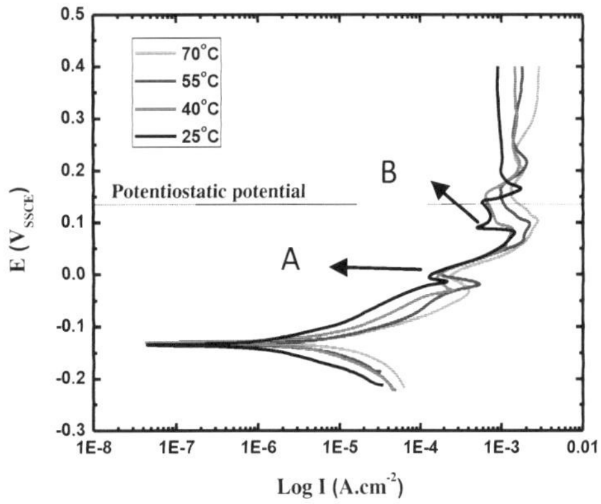

Figure 1. Potentiodynamic polarization of silver-copper-tin brazing alloy

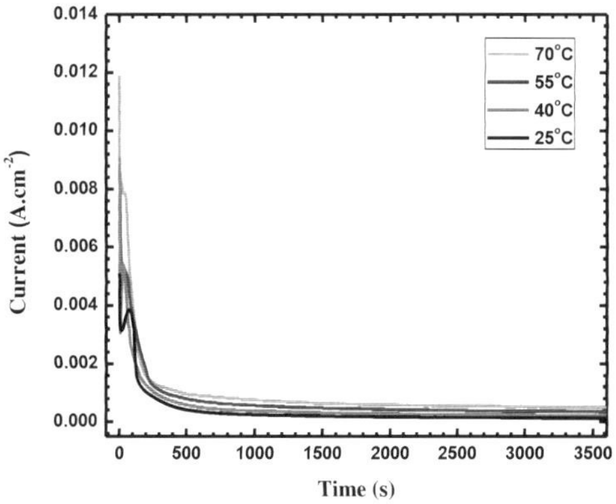

Figure 2. Potentiostatic polarization of silver-copper-tin brazing alloy with the potential of 135 mV_{SSCE} at 25, 40, 55, and 70°C

At higher temperature the E_{corr} tend to shift towards more noble value with respect to the increased of temperature. The E_{corr} at 40°C, 55°C, and 70°C are -0.130 V_{SSCE}, -0.134 V_{SSCE}, -0.124 V_{SSCE} respectively. Interestingly, the corrosion current showed the same manner regarding its relationship with the changes of temperature which is tends to increase with the increasing of temperature: 9.72×10^{-7} A at 40°C, 1.33×10^{-6} at 55°C, and 1.36×10^{-6} at 70°C.

It is believed that the m A ls might consists layer of corrosion products that formed during the short preliminary hold time at the free corrosion potential prior the tests [5, 6]. As has been known, temperature favours the kinetics of the reactions, and supposed to have the similar effect with these corrosion products layer forming reactions. Thus, the E_{corr} was slightly shifted to the noble values with increasing of temperature. However, since temperature also tends to increase the kinetic of the anodic dissolution reactions, the anodic current densities are higher as temperature increases.

At the E_{corr} corrosion possibly associated with the dissolution of Sn to Sn(II) species [7, 9, 10] based on the following reactions:

$$Sn + 2OH^- = Sn(OH)_2 + 2e^- \quad (1)$$

$$\text{and/or } Sn + 2OH^- = SnO + H_2O + 2e^- \quad (2)$$

$$Sn(OH)_2 = SnO + H_2O \quad (3)$$

The passivation of the alloy in the region of the potential slightly above the E_{corr} (denoted as A in figure 1) possibly due to the formation of Sn(IV) species by the following reactions:

$$SnO + H_2O + 2OH^- = Sn(OH)_4 + 2e^- \quad (4)$$

$$Sn(OH)_2 + 2OH^- = Sn(OH)_4 + 2e^- \quad (5)$$

The $Sn(OH)_4$ is insoluble and precipitated on the alloy's surface forming a passivating film. However, chlorination of the corrosion product may occur and the chlorides have incorporated in the lattice of Sn(IV) species. Furthermore, the copper might also form CuCl mixed with the tin rich corrosion product [11, 12]. The Sn(IV) film dehydrates to form SnO_2, and resulted decreasing of the current density at the upper region (denoted as B in figure 1) of the potentiodynamic polarization curves particularly at 25°C. These passive films tend to become less stable with increasing of the temperature. At the highest temperature (70°C) the potentiodynamic polarization curve shows that the anodic current density continuously increases.

The potentiostatic experiments were done on purpose to evaluate the passive region as a function of temperature. The curves represented in figure 2 showed that the current resulted from increasing the temperature in the passive region become greater, even so the differences are not large. The current curves showed the highest degree of stability. The potentiostatic curves were also shown no oscillation or abruption, hence it is possible that the alloy was in the passive state with relatively stable passive layer at each of the temperature. Despite the insoluble mixed corrosion product of Sn and Cu, silver as the noblest element in the alloy is seems to have an important contribution to the low current density, either it is in its origin or its oxide form.

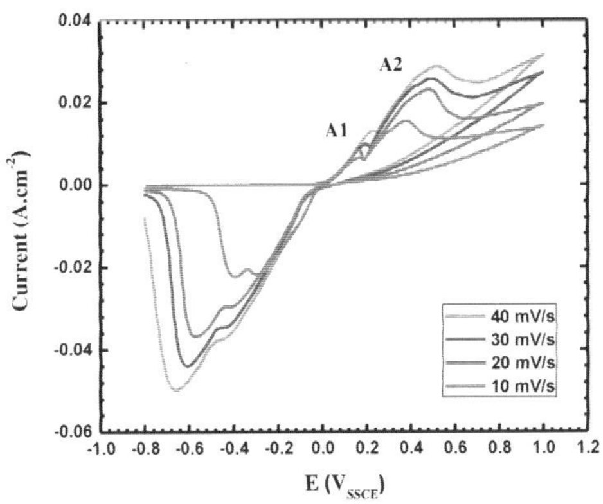

Figure 3. Cyclic voltammogram of silver-copper-tin brazing alloy in NaCl 3.5 wt% at 10, 20, 30, 40 mV/s

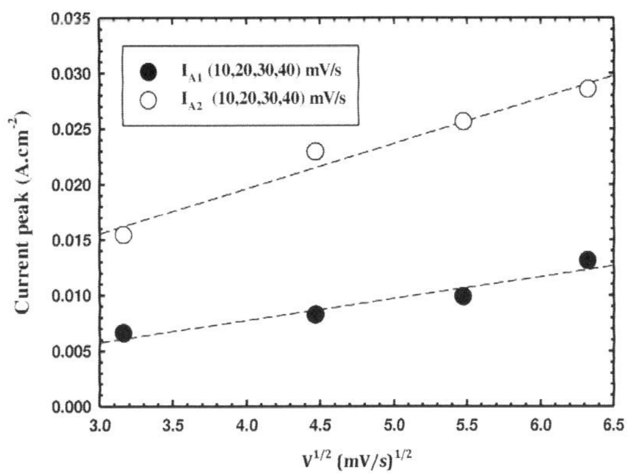

Figure 4. Silver-copper-tin brazing alloy anodic peak current versus square root of scan rate at 10, 20, 30, 40 mV/s

Voltammograms of the silver-copper-tin brazing alloys in figure 3 showed that the cathodic peak is obviously appeared at any scan rate. Furthermore, the peak increasing in current value along with scan rate. Two peaks of the anodic part are shown and noted as A1, and A2. The A1 peak is well-defined at the small scan rate, then gradually reduced and more like shoulder with increasing scan rate. The dependence of Ea and Ia on the v 1/2 in figure 4 showed a high degree or linearity. The linear relationship between Ia on the v 1/2 described that the anodic reaction was diffusion controlled by the limiting diffusion species. From the report by M. Drogowska et al. [10] the diffusing species differ in nature from the ions already present in solution as OH^-, the possibly limited diffusing species is Sn^{2+}. The A1 peak indicated the diffusion-controlled of tin and, subsequently, the formation of porous oxides or hydroxides. Relied on the potentiodynamic polarization curves, good correlation between a small peak near the corrosion potential of the alloy and the first peak (A1) of the alloy's voltammograms established that the production of chlorinated and porous corrosion product occured in this region

The corrosion of the alloy initiated by the dissolution of Sn to the Sn(II) species as proposed by several authors [26-31]. The dissolution of Sn to form Sn(II) species predictically could inhibit the corrosion to progresses. In accordance with the potentiodynamic polarization result, the $Sn(OH)_2$ and SnO firstly formed and followed by the production of relatively insoluble $Sn(OH)_4$ and chlorinated corrosion product (possibly $SnO_x(OH)_yCl_z . nH_2O$, with $z \leq 1$ and CuCl) before finally resulted the more stable product of SnO_2 that makes the curves decay before reaches relatively steady state at lower current density after slightly increasing at the initial. Since the static polarization experiments have been done at passive region of the alloy (second peak of the minimum current density), the stable passive film of SnO_2 possibly achieved.

4. Conclusion

The dissolution of Sn seems to be initiating the corrosion of the silver-copper-tin brazing alloy. Nevertheless, the active dissolution of the Cu simultaneously influences the corrosion behavior of the silver-copper-tin brazing alloy. A typical anodic dissolution of Cu-Sn metal alloys showed by the alloy. At least two regions of the active-passive transition are observed. These minimum peaks of the current density probably caused by the formation of mixed tin rich insoluble corrosion product and copper corrosion product. The alloy was in the stable passive state regardless the temperature variably.

5. References

[1] Sigler D R, Schroth J G, Wang Y, Radovic D 2007 *Weld. J.* **86** 340
[2] Joseph A W, Jorge M M 2008 *GE Power Systems*
[3] Kawakatsu I 1973 *Weld. J.* **20** 233
[4] American Welding Society 1991 *Brazing Handbook* 4th Edition (Miami: United States) 284
[5] Udit S M and Kwang-Lung L 2006 *Corros. Sci.* **48** 662
[6] Oulfajrite H, Sabbar A, Boulghallat M, Jouaiti A, Lbibb R, Zrineh A 2003 *Mater. Lett.* **57** 4368
[7] Pugh M, Warner L M, Gabe D R 1967 *Corros. Sci.* **7** 807
[8] Dickinson T and Lotfi S 1978 *Electrochim. Acta* **23** 513
[9] Stirrup B N and Hampson N A 1977 *Surface Technology* **5** 429
[10] Drogowska M, Ménard H, Brossard L 1991 *J. Appl. Electrochem* **21** 84.
[11] Robbiola L, Tran T T M, Dubot P, Majerus O, Rahmouni 2008 *Corros. Sci.* **50** 2205
[12] Tsao L C and Chen C W 2012 *Corros. Sci* **63** 393

IOP Conf. Series: Materials Science and Engineering **526** (2019) 012034 doi:10.1088/1757-899X/526/1/012034

The Performance of Permanent Magnet Core of Reducing Sag Voltage in Personal Computer (PC)

H Ys *, A D Simanjuntak
Electrical Engineering Department, Universitas Kristen Maranatha, Bandung, Indonesia.

*Corresponding author: siti_herawati_aminah@yahoo.com

Abstract. The power quality has a significant effect on all system devices. The voltage factors are usually modulated caused by the excessive load changes which produce a sag voltage. One of the techniques used to reduce the voltage sag is to design a Cuk converter with PI controller coupled with two inductances (L) at the Cuk converter in order to make a non-ideal transformer with a permanent magnet core made by ferrite material. The experimental results revealed that, the non-ideal transformer with a permanent magnetic core would stabilizes the input voltage on personal computer (PC), thus the sag voltage can be minimalized. From the experiment, the optimum results were obtained with top gap was 1mm and the bottom gap was 6mm. It can be concluded that by using a ferrite core magnet with high coercivity, the output voltage can be stabilized thus there will be no sag at the input voltage on the PC.

1. Introduction

The power interruption is often occurred in electrical systems, the sensitivity of modern electronic devices today is more susceptible to the quality of the power supply [1]. Power quality is more related to the ability of the electrical system to provide uninterrupted power. However, there are some power quality problems such as sag, swell, harmonic distortion, imbalance, flicker and transients. Among them, sag voltage occurs more often compared to other problems [2]. The power sag quality problems arise about 60% larger than the swell voltage (29%), transient (8%) and disruption problem (3%) [1]. Although the duration time of the sag voltage is short (from 0.5 to 1 minute) but gives impact to the operation of the equipment (motor, trip on sensitive load and inaccurate control equipment) [3]. In addition, the instantaneous sag voltage can cause serious problems with computer systems or electronic equipment [4].

The power system demands a reliable electrical system. The system is said to be reliable if it continues to supply an electricity need, with a constant voltage and frequency thus meet the consumers' urgency. If there is a tension sag then the voltage becomes unstable, the swell voltage should be muted in order to make a stable voltage. In this study, the design of permanent magnet core was done to stabilize the electrical voltage with the standard input voltage personal computer. The design will be simulated using PSIM (power sim).

2. Perfomance of Permanent Magnet Core of Reduce Sag Voltage

Referred to IEEE-1559-1995 standard, sag voltage is defined as a sudden reduction of RMS value from 10-90% at nominal voltage for duration of 0.5 to 1 minute [5]. Specific ranges of input power

IOP Conf. Series: Materials Science and Engineering **526** (2019) 012034 doi:10.1088/1757-899X/526/1/012034

quality and load parameters for computer manufacturing are voltage distortion parameters (all phases) if sag occurs then the range is -18% for a maximum of 0.5 s [6]. Cuk converter can raise and lower the voltage, so it can work on various input voltage range. This input voltage modulation is done by varying the duty cycle, since the duty cycle will keep the output voltage constant [7]. In designing the Cuk converter circuit, 3 modification stages were simulated which were simulated using PSim (Power Simulation) software. The first stage is to design a closed loop converter loop circuit using the PI controller. From the Cuk converter circuit the following formula is obtained:

$$Duty\ cycle = -\frac{Vo}{Vs-Vo} \tag{1}$$

Simulation of the Cuk converter is done by changing the parameter values of Li, Lo, Ci and Co [8], then we get one example of output voltage in Cuk converter circuit when: V_{in} = 220 V, Li = 56 mH, Lo = 58.5mH, Ci = 1e-4F, Co = 1e-5 F as shown in Figure 1.

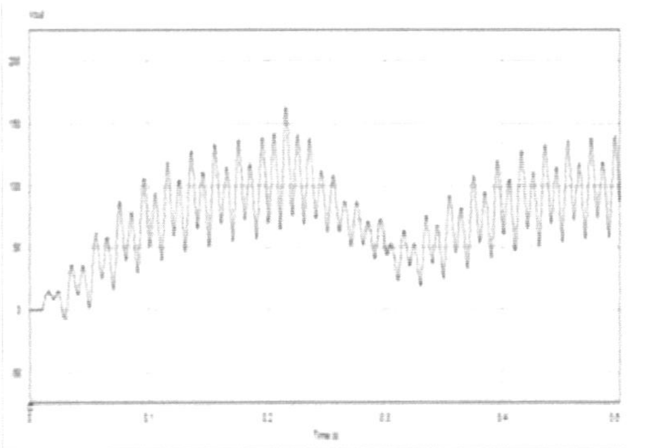

Figure 1. Cuk Converter Output Voltage Results

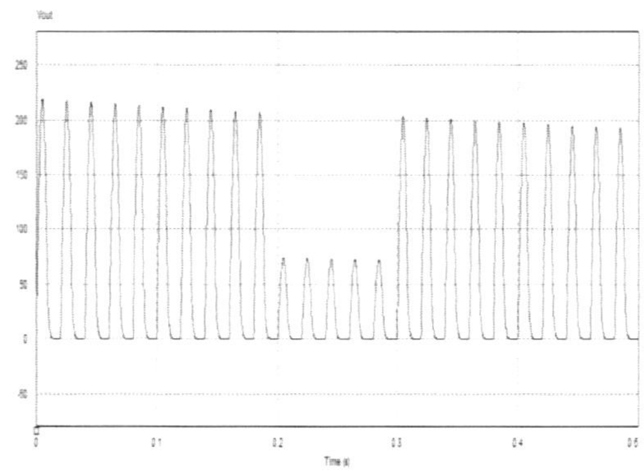

Figure 2. Output Results Using Ideal Transformer

In the figure 1 we can see that the output voltage is unstable with a peak value of 165 V. Because the sag voltage converter circuit has not been able to be muted, the next step is to couple the two inductance (L) on the Cuk converter circuit becomes the ideal transformer. The ideal transformer uses 5 primary windings and 5 secondary windings. Output results using ideal transformer are shown in Figure 2. It can be seen in figure 2 that the output voltage is still not stable and at 0.2 s - 0.3 s there are also some voltage sag. Then the last step taken is the transformer is not ideal to use a magnetic core. The permanent magnet can be a current source and control the voltage source [9]. In this non-ideal transformer, the important values that have to be known are copper wire resistance and the cross-sectional area of the magnetic air gap [10]. The air gap is given a range of 1-10 mm, and the saturation value of the hysteresis curve corresponds to the hysteresis curve of the ferrite magnet. The crane of the converter converter with the non-ideal transformer is shown in Figure. 3.

2.1. Air Gap Magnet

The magnetic core used is type-E which has 6 air gaps; 3 on the top and 3 on the bottom. The cross-sectional area of each air gap in the magnet is shown in Table 1. The air gap is used in the range of 1-10 mm.

IOP Conf. Series: Materials Science and Engineering **526** (2019) 012034 doi:10.1088/1757-899X/526/1/012034

Table 1. Sectional Area of the Air Gap

	Cross-sectional area (m^2)		
Top Gap	Gap 1	Gap 2	Gap 3
	7.04E-05	1.33E-04	7.04E-05
Under Gap	Gap 4	Gap 5	Gap 6
	7.04E-05	1.33E-04	7.04E-05

2.2. Ferrite Magnetic Hysteresis Curve

B-H curve which resulted from the PSim software simulation on the ferrite magnet was shown in Figure 4. Ferrite magnets have high coercivity with stability level against external field influence and good temperature [11].

Figure 3. Permanent Magnet Shaped E

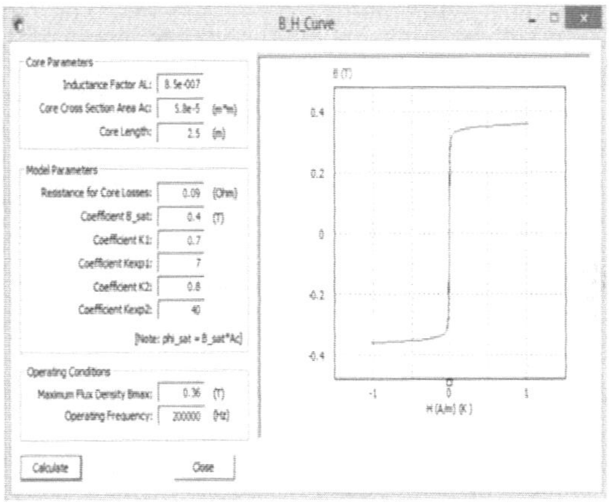

Figure 4. B-H Curve

From Figure 4, there are parameter values. These parameter values are then fed into the core component of PSim.

2.3 Resistance (R) on Copper

The resistivity (ρ) of copper is used following the conductor material type. The purpose of using R value in the copper wire is because there are losses on the copper wire which will show on how the result a non-ideal transformer. Value resistance (R) on copper is 0.011 Ω.

2.4 The output voltage of non-ideal transformer with a magnetic Core

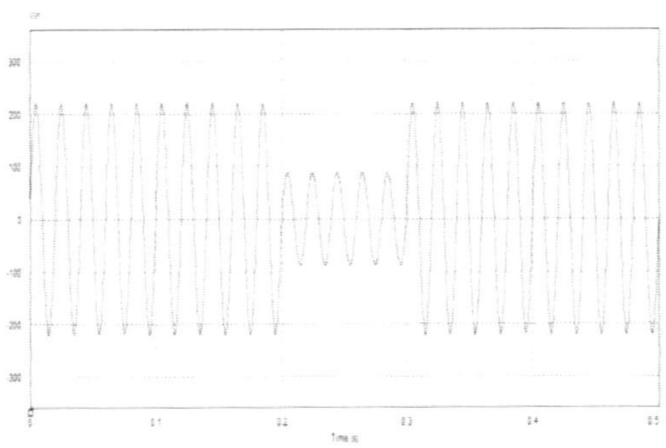

Figure 5. Input Voltage Sag

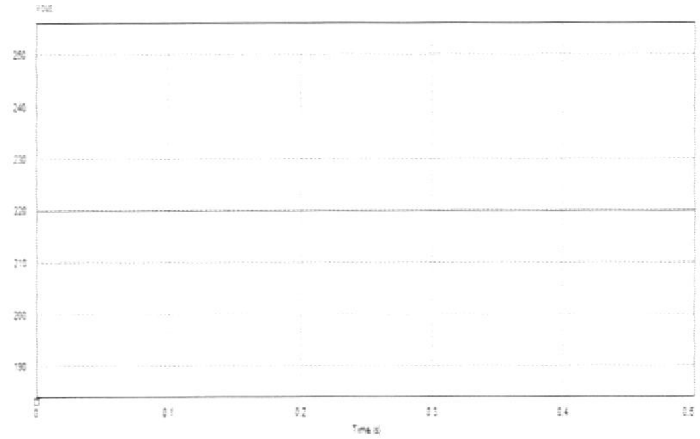

Figure 6. Output Voltage

The input voltage power supply on the PC still has a sag problem as shown Figure 5. However, after sag damping using magnetic core, the input voltage on the PC becomes stable at 220 V. The output voltage of the ideal transformer with the magnetic core is shown in Figure 6.

3. Conclusions

From result and discussion, it is revealed that:

1. The initial sag voltage of 60% of 220V, can be reduced to 0%. In other word the sag voltage is totally lost.
2. Sag voltage can occur due to the voltage drop at the input voltage. This will affect an unstable voltage which will damage computer equipment especially the one with power supply. A transformer with a high permeability magnet core is able to stabilize the input voltage, thus the sag voltage can be reduction.
3. From the experimental results, the best result was obtained when the top gap = 1mm and the bottom gap = 6mm.

4. References

[1] M. Ravilla and R. R. G, "Modeling and Simulation of a Distribution STATCOM (D-STATCOM) for Power Quality Problems-Voltage Sag and Swell Based on Sinusoidal Pulse Widht Modulation (SPWM)," *IEEE-International Conf. Adv. Eng. Manag.*, pp. 436– 441, 2012.

[2] K. Yamamoto, S. Ehira, M. Ikeda, S. I. Hase, and M. A. C.Onverter, "Synchronous Frame Control for Voltage Sag / Swell Compensator Utilizing Single-Phase Matrix Converter," *IEEE*, 2017.

[3] D. V. Tien and R. Gono, "Analysis and Simulation of Causes of Voltage Sags Using EMTP," *IEEE*, no. **4**, pp. 7–11, 2017.

[4] U. V. Reddy, P. C. Babu, and S. S. Dash, "Space Vector Pulse Width Modulation Based DVR to Mitigate Voltage Sag and Swell,"*IEEE*, pp. 5–9, 2013.

[5] A. Tavighi, H. Abdollahzadeh, and J. Martí, "Fast Response DVR Control Strategy Design to Compensate Unbalanced Voltage Sags and Swells in Distribution Systems," *IEEE*, 2013.

[6] P. Systems and E. Committee, *IEEE Recommended Practice for Emergency and Standby Power Systems for Industrial and Commercial Applications*, vol. 1995. 1995.

[7] A. Triandini, Soeprapto, and R. Mochammad, "Perancangan Battery Control Unit (BCU) Dengan Menggunakan Topologi Cuk Converter Pada Instalasi Tenaga Surya," pp. 1–6, 2013.

[8] Rheesabh Dwivedi, Vinay Kumar Dwivedi, and Rahul Sharma, "Parametric variation analysis Of CUK converter for constant voltage applications", *IJAREEIE*, Vol. 3, Issue 2, February 2014.

[9] H. Yusuf, E. T. Raharjo, and I. Ganiwa, "Prototipe Magnet Ferrit untuk Mengendalikan Akumulasi Tegangan Cuk Konverter," *J. Elektron. dan Telekomun.*, vol. **9**, no. 2, pp. 209–212, 2009.

[10] David M. Pozar. 2012. "*Microwave Engineering, Fourth Edition*".

[11] N. A. Zen, W. Widanarto, and T. Cahyanto, "Karakterisasi Struktur dan Sifat Magnetik Manganese Ferrite sebagai Bahan Magnet Permanen Isotropik," in *Prosoding Pertemuan Ilmiah XXVIII HFI*, 2014, no. April, pp. 268–271.

Development of tyre tread compounds for good wet-grip : effects of rubber type

K Suchiva[1,*], C Sirisinha[1], P Sae-oui[2], and P Thapthong[2]

[1] Rubber Technology Research Centre, Faculty of Science, Mahidol University, Nakhon Pathom 73170, Thailand
[2] National Metal and Materials Technology Center, National Science and Technology Development Agency, Pathum Thani 12120, Thailand

*Corresponding author: Krisda.suc@mahidol.ac.th

Abstract. The present work reported the study of the effects of type of rubber on energy loss of tyre tread compounds for radial tyres which determines wet-grip properties of the tyres derived from them. The types of rubbers studied were solution SBR (S-SBR), emulsion SBR (E-SBR), functionalized S-SBR (FS-SBR), natural rubber (NR), functionalized NR (CNR), butadiene rubber (BR) and their selected blends.The wet-grip property was assessed by measuring tan δ at 0°C. It was found that FS-SBR gave tread compound with the best wet-grip property followed by S- SBR, CNR, E-SBR, NR and BR. Blends of SBR with lower ranking rubbers (CNR, NR and BR) also gave lower values of tan δ at 0°C or poorer wet-grip properties. The results can be explained by the viscoelastic properties of the rubbers and their interactions with silica.

1. Introduction

The trend in development of tyres are towards low rolling-resistance, good wet-grip and low noise tyres. The present research was undertaken with the objective to obtain knowledge and technologies for production of energy-saving, good wet-grip and low noise tyres and disseminates to the Thai tyre manufacturers.

In this research work the effects of rubber type (emulsion SBR, solution SBR, NR, BR, functionalized solution SBR, chloro-acetated NR) and their blends, reinforcing filler type (carbon black, silica) and their mixtures, type of sulphur vulcanisation system (CV, semi-EV, EV), crosslink density, functionalised plasticizer (MA-g-PB) and mixing conditions on energy dissipation of tyre tread compounds, which determines wet-grip properties of the tyres derived from them, were studied. However, only the effects of rubber type and selected blends will be reported here. The compound formulations were based on silica and conventional sulphur vulcanisation system which were found to be the best reinforcing filler and vulcanisation system for wet-grip property. [1] Use of hybrid fillers (silica and carbon black) were also studied but the results are reported elsewhere. [2]

2. Experimental

The compound formulation used in this study is shown in Table 1. Mixing was carried out by using a laboratory internal mixer (Brabender-Plasticorder 350E). Mixing was done in 3 steps, (1) all ingredients except curatives were mixed using the fill factor of 0.75, initial chamber temperature of 60°C, rotor speed of 40 rpm and mixing time of 10 minutes (2) the obtained compounds were mixed at 140°C, for 6 minutes to ensure complete reactions between rubber, silane coupling agent (Si-69) and silica (3) the compounds were finally mixed with curatives for 3 minutes. The optimum cure time (tc90) was determined using a moving die rheometer (TechPro MD+) at 160°C. Rubber-filler interaction was evaluated through the measurement of bound rubber content (BRC) by immersing approximately 0.5 g of the compound in 100 ml toluene at room temperature for 7 days. After filtration, the rubber-filler gel was completely dried and weighed. The BRC was then calculated using Equation (1).

$$\%BRC = \frac{W_{fg} - WF_f}{WF_p} \times 100 \tag{1}$$

where W_{fg} is the weight of rubber-filler gel, W is the weight of the test specimen. Ff and Fp are the weight fractions of filler and polymer in the rubber compound, respectively.

The tensile properties were determined by using a universal testing machine (Instron 3366, USA). Dependence of loss factor (tan δ) on temperature was measured by dynamic mechanical analyzer (Gabo, Eplexor 25N, Germany) in tension mode. Temperature was scanned from -60 to 80 °C at 2/min under static strain, dynamic strain and frequency of 1%, 0.15% and 10 Hz, respectively.

Table 1. Compound formulation for the study of effects of rubber type

Materials	Amount (part by weight)	
SSBR or E-SBR or F-SSBR or NR or CNR or BR	100	60
NR or CNR or BR	0	40
ZnO	3	3
Stearic acid	2	2
6PPD	1.5	1.5
TMQ	1	1
Paraffin wax	2	2
Silica	48	48
TESPT (10% w/w of silica)	4.8	4.8
TDAE oil	3	10
TBBS	1.2	1.2
TBzTD	0.2	0.2
Sulphur	2.2	2.2

3. Results and Discussion

Table 2 summarises results of tensile measurements for various vulcanised rubbers studied.

Table 2. Tensile properties of the rubber vulcanisates studied

Rubber type	10% Modulus (MPa)	100% Modulus (MPa)	300% Modulus (MPa)	Tensile strength (MPa)	Elongation at break (%)
SSBR	0.65	3.45	14.4	19.5	389
FS-SBR	0.68	4.26	18.8	23.1	362
E-SBR	-	2.20	-	19.3	499
NR	0.83	4.51	17.9	24.7	433
CNR	0.84	5.20	19.5	23.2	368
BR	-	1.73	-	13.2	511
S-SBR/NR (60/40)	0.74	3.60	13.9	19.6	413
S-SBR/CNR (60/40)	0.64	3.68	15.2	18.4	357
S-SBR/BR (70/30)	-	2.38	-	16.1	414
E-SBR/NR (70/30)	-	2.01	-	16.3	515
E-SBR/BR (70/30)	-	2.17	-	16.8	479
F-SSBR/NR (60/40)	0.76	4.00	16	20	369
F-SSBR/CNR (60/40)	0.72	5.51	19.2	19	292

Results of bound rubber measurement are presented in Figure 1.

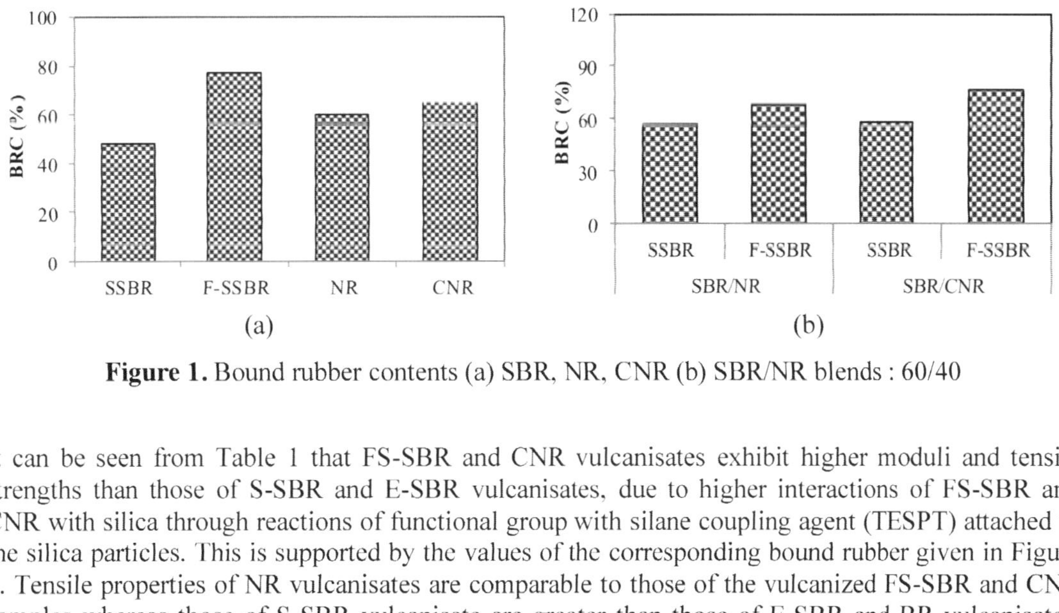

Figure 1. Bound rubber contents (a) SBR, NR, CNR (b) SBR/NR blends : 60/40

It can be seen from Table 1 that FS-SBR and CNR vulcanisates exhibit higher moduli and tensile strengths than those of S-SBR and E-SBR vulcanisates, due to higher interactions of FS-SBR and CNR with silica through reactions of functional group with silane coupling agent (TESPT) attached to the silica particles. This is supported by the values of the corresponding bound rubber given in Figure 1. Tensile properties of NR vulcanisates are comparable to those of the vulcanized FS-SBR and CNR samples whereas those of S-SBR vulcanisate are greater than those of E-SBR and BR vulcanisates. The moduli and tensile strengths of FS-SBR, S-SBR and E-SBR blended with NR or BR show lower values compared with the corresponding unblended rubbers.

The energy loss of the rubber vulcanisate or tan δ at 0°C, as measured by dynamic mechanical method, is used as a measure of good wet-grip properties. High value of tan δ at 0°C implies good wet-grip property. Results of tan δ measurement at 0°C of various vulcanised rubber compounds are shown in Table 3. It can be observed that FS-SBR give vulcanised tread compound having the best wet-grip property followed by S- SBR, CNR, E-SBR, NR and BR. Blends of SBR with lower ranking rubbers (CNR, NR and BR) also gave lower values of tan δ at 0°C or poorer wet-grip properties.

Table 3. Values of tan δ at 0oC of various rubber vulcanisates studied

Rubber Type	Blend Ratio	tan δ at 0°C
FS-SBR	-	0.726
S-SBR	-	0.538
E-SBR	-	0.237
CNR	-	0.307
NR	-	0.183
BR	-	0.147
FS-SBR/CNR	60/40	0.552
FS-SBR/NR	60/40	0.493
S-SBR/CNR	60/40	0.527
S-SBR/NR	70/30	0.458
E-SBR/NR	70/30	0.231
S-SBR/BR	70/30	0.286
E-SBR/BR	70/30	0.147

4. Conclusion

From the study it was found that FS-SBR gave tyre tread compound with the best wet-grip property followed by S- SBR, CNR, E-SBR, NR and BR, as revealed by the values of tan δ at 0°C. Blends of SBR with lower ranking rubbers (CNR, NR and BR) also gave lower values of tan δ at 0°C or poorer wet-skid resistance.

5. References

[1] P. Sae-oui, K. Suchiva, C.Sirisinha,W. Intiya, P.Yodjun and U. Thepsuwan, Effects of Blend Ratio and SBR Type on Properties of Carbon Black-Filled and Silica-Filled SBR/BR Tire Tread Compounds, Advances in Materials Science and Engineering, Article ID 2476101 (2017)

[2] P.Thaptong,P.Sae-Oui,C.Sirisinha,InfluencesofStyrenebutadienerubberandSilicatypeson Perfor mance of Passenger car radial tire tread, Rubber Chemistry and Technology, 90(4):699-713 (2017)

Carbon sequestration potential of rubber-tree plantation in Thailand

D Satakhun[1], C Chayawat[1], J Sathornkich[2], J Phattaralerphong[3], P Chantuma[4], P Thaler[5], F Gay[5], Y Nouvellon[5], and P. Kasemsap[2,*]

[1] Center of Thai-French Cooperation on Higher Education and Research, Kasetsart University, Bangkok 10900, Thailand.
[2] Department of Horticulture, Faculty of Agriculture, Kasetsart University, Bangkok 10900, Thailand.
[3] Agro-Bioresources Program, Faculty of Natural Resources and Agro-Industry, Kasetsart University, Sakon Nakhon 47000, Thailand.
[4] Chachoengsao Rubber Research Center, Chachoengsao 24160, Thailand
[5] CIRAD, UMR Eco&Sols, F-34060 Montpellier, France.

* Corresponding author: agrppk@ku.ac.th

Abstract. Thailand has approximately 3 million hectares of latex-producing natural rubber plantations. These rubber plantations have large potential to sequester atmospheric carbon into the biomass and soil. We use eddy covariance technique to measure the net ecosystem exchange (NEE), which is the vertical net flux of CO_2 above the rubber plantation, as the estimate of CO_2 sequestration potential of rubber plantation. NEE of a 19-year old, latex-producing rubber plantation of a monoclonal stand of rubber trees (*Hevea brasiliensis* Müll.Arg.) clone RRIM 600 at Chachoengsao Rubber Research Center was observed for 4 years, starting from 2013. The results showed that annual CO_2 sequestration from 2013 to 2016 ranged from 28.0 to 43.1 tons CO_2 ha^{-1} yr^{-1} and averaged 36.7 tons CO_2 ha^{-1} yr^{-1}. In 2016, there was 2.95 million hectares of latex-producing natural rubber plantations in Thailand producing 4.342 million tons of natural-rubber latex. Thus, the estimated CO_2 sequestration potential of these plantations was approximately 108 million tons of CO_2. In addition, it was estimated that rubber plantation in Thailand sequestered approximately 24.9 kg of CO_2 to produce each kilogram of natural-rubber latex. This large carbon sequestration potential associated with natural latex production information shows that natural latex is very friendly to the environment. These observations have important implications not only for the potential management of carbon sequestration in Thailand, but also for the establishment of marketing strategy to enhance the natural rubber competitiveness, especially in comparison to the less environmental-friendly synthetic rubber.

1. Introduction

Rubber (*Hevea brasiliensis* Müll. Arg.) is one of the most economically important tree crops in tropical area throughout the world. Rubber plantations provide natural rubber latex and rubber wood for several downstream industrials. There are approximately 3 million hectares of latex-producing natural rubber plantations in Thailand, providing primary income for approximately 6 million people.

In addition to providing latex and wood, these rubber plantations have large potential to sequester atmospheric carbon into the biomass and soil [6], [7], [8], [11], [12], [13]. However, the potential role of these plantations in sequestering carbon in the soil and plant biomass in relations to latex produced, which could be the most important indicator to illustrate the environmental-friendliness of natural rubber latex, has not been fully evaluated.

The vertical net flux of CO_2 above the ecosystem, often called the net ecosystem exchange (NEE), includes net ecosystem production (NEP) and inorganic sources and sinks for CO_2. NEP, which is defined as the difference between the organic carbon (C) fixed by photosynthesis in an ecosystem (gross primary production, or GPP) and released by total ecosystem respiration (the sum of autotrophic and heterotrophic respiration, or R_e), represents organic C available for storage within the system or loss from it by export or non-biological oxidation [5]. Inorganic sources and sinks, such as weathering reactions, precipitation or dissolution of carbonates, and atmosphere–water equilibrations, are likely to be minor terms in the CO_2 exchange such as in a forest ecosystem [5]. Eddy Covariance (EC) technique has become one of the preferred micrometeorological methods for measuring the CO_2, water vapor, and energy exchange between the atmosphere and terrestrial ecosystems. The EC technique is now widely used to measure the vertical turbulent CO_2 flux between the atmosphere and biosphere as it provides continuous flux information integrated at the ecosystem scale, and annual carbon sink or source strength for a wide variety of ecosystems [1], [2], [9], [10]. In recent years, many studies have used EC techniques to measure NEE [3], [4], [10].

The present study addresses two main objectives: to estimate CO_2 sequestration potential of Thai rubber plantations by NEE measured using eddy covariance technique, and to quantify how much carbon dioxide is sequestered by Thai rubber plantation to produce a kilogram of natural-rubber latex.

2. Materials and methods

The research was done at the Chachoengsao Rubber Research Center located in Chachoengsao province (13°41′N, 101°04′E), Thailand. The observation site is a monoclonal stand of rubber trees clone RRIM 600. The rubber trees were 19-year-old in 2013. Tapping for latex production began when the trees were 8-year-old. The Eddy Covariance (EC) instruments were mounted at a height of 25 m on a tower. EC system consisting of a three-dimensional ultrasonic anemometer (CSAT-3, Campbell Scientific Inc., USA.) and the open-path infrared gas analyser (LI-7500, LI-COR Inc., USA) which were used to measure speed and direction of the wind and concentration of CO_2 and water vapour, respectively. An assumption of the EC technique is the flux of a given scalar parameter can be measured as an average of the covariance between the 20-Hz fluctuations in the vertical wind speed and the 20-Hz fluctuations of the scalar parameters. CO_2 flux was calculated as the mean covariance of vertical wind velocity and scalar fluctuations, with the appropriate corrections applied. The following equation presents the calculation of CO_2 flux:

$$F_{CO_2} = \overline{\rho_a} \cdot \overline{w'c'} \tag{1}$$

Where (ρ_a) is a density of the air, w' and c' are vertical wind speed and CO_2 concentration fluctuations from the means, respectively. Overbars in the equation show time averaging and primes represent fluctuations from the mean value. All raw data were collected at a rate of 20 Hz by a data logger (CR3000, Campbell Scientific Inc., USA). The 30-min mean CO_2 fluxes were calculated by using all raw data. Before covariance calculation, the time series data were de-spiking and linearly detrended. The fluxes were three-dimensional coordinate rotations to align the sonic anemometer axis along the long-term streamlines and WPL-correction. Following the sign convention in the atmospheric flux community, negative flux covariance indicates a loss of carbon from the atmosphere and gain by the ecosystem. CO_2 sequestration potential of rubber plantation was taken to be equal to –NEE, in order to indicate the positive annual net carbon gain by the rubber ecosystem.

Along with the EC measurements, standard meteorological data and soil parameters were measured continuously with an array of sensors. Quantum sensor (LI-190SB, LI-COR Inc., USA) was used to measure photosynthetically active photon flux (PPF). Solar radiation and net radiation were measured by using sunshine sensor (BF3, Delta-T Device Ltd., UK) and net radiometer (NR-Lite, Kipp & Zonen, The Netherlands), respectively. Rainfall was measured by using the tipping bucket rain gauge (TE525, Campbell Scientific Inc., USA). Air temperature and relative humidity were measured by the temperature and relative humidity probe (HMP45C, Vaisala, Finland). Belowground parameters were measured on 3 locations in the rubber plantation. Soil temperature at depths of 1 m from soil surface was measured with a custom built chromel-constantan thermocouple. Soil volumetric water content at depths of 1 m. from soil surface was measured with Water Content Reflectometer (CS616, Campbell Scientific Inc., USA.). All parameters were measured every 10s and the 30 min averaged values were recorded in data logger (CR1000, Campbell. Scientific Inc., USA.).

3. Results and discussion

From 2013 - 2106, minimum, maximum, and mean values of air temperature ranged from 12.9-15.6, 37.9-39.9, and 27.2-27.8 °C, respectively (Table 1). Soil temperature at 100 cm below surface ranged from 25.9 to 26.5°C while soil water content (SWC) at 100 cm depth ranged from 0.141 to 0.181 m^3 m^{-3}. The annual averaged relative humidity was greater than 80%. The annual cumulative rainfall showed much greater variability (24.4%), ranging from 1,023 to 1,316 mm yr^{-1}, than PPF (5.4%), solar radiation (4.7%) or net radiation (12.3%).

In this study -NEE, taken as proxy for annual CO_2 sequestration, ranged from 28.0 to 43.1 tons CO_2 $ha^{-1}yr^{-1}$ and averaged 36.7 tons CO_2 ha^{-1} yr^{-1} (Table 2). Simple linear regression analysis (not shown) indicated that PPF, the major driver for photosynthesis, was positively correlated to annual CO_2 sequestration and explained 81.5% of its variations. In addition, annual CO_2 sequestration was positively related to annual latex yield and explained 61.5% of its variation (Figure 1).

In 2016, there was 2.95 million hectares of latex-producing natural rubber plantations in Thailand producing 4.342 million tons of natural-rubber latex. Thus, the estimated CO_2 sequestration potential of these plantations was approximately 108 million tons of CO_2. In addition, it could be estimated further that rubber plantation in Thailand has the potential to sequester approximately 24.9 kg of CO_2 to produce each kilogram of natural-rubber latex. On the other hand, the inverted slope of the linear regression function between annual CO_2 sequestration and annual latex yield in Figure 1 indicates that an increase of 22.1 kg of sequestered CO_2 is correlated to an increase of 1 kg latex yield. This large carbon sequestration potential, 22.1 to 24.9 kg CO_2 sequestered per 1 kg of natural-rubber latex production, shows that natural rubber is very friendly to the environment especially when compared to the net carbon released associated with synthetic rubber production. These observations have important implications not only for the potential management of carbon sequestration in Thailand, but also for the establishment of marketing strategy to enhance the natural rubber competitiveness, especially in comparison to the less environmental-friendly synthetic rubber.

Table 1 Annual averaged temperatures, soil water content, and relative humidity and accumulated rainfall, photosynthetically active photon fluxes, solar radiation and net radiation at Chachoengsao Rubber Research Center.

Year	2013	2014	2015	2016	Mean
Air temperature (°C)					
minimum	14.3	12.9	15.6	13.3	14.0
maximum	38.3	37.9	38.2	39.9	38.6
average	27.2	27.3	27.7	27.8	27.5
Soil temperature (°C) at 100 cm. depth	26.4	25.9	26.0	26.5	26.2
SWC (m^3 m^{-3}) at 100 cm. depth	0.181	0.179	0.163	0.141	0.166
Relative humidity (%)	81.7	81.5	80.8	81.4	81.4
Rainfall (mm yr^{-1})	1,316	1,282	1,161	1,023	1,196
PPF (mol photon m^{-2} yr^{-1})	5,418	5,725	5,688	5,665	5,624
R_g (MW.m^{-2} yr^{-1})	3,169	3,314	3,258	3,321	3,266
R_n (MW.m^{-2} yr^{-1})	2,163	2,208	1,952	1,992	2,079

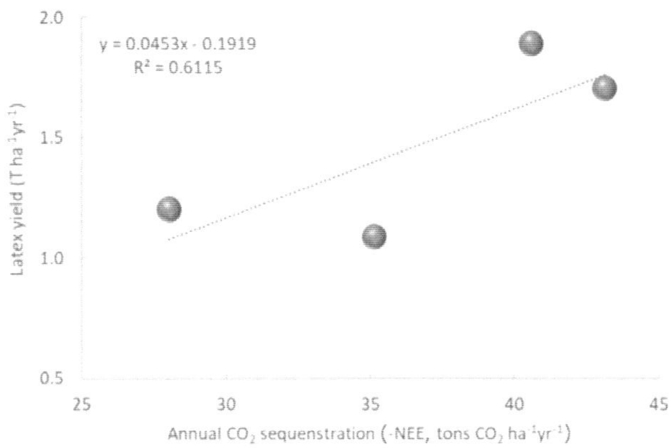

Figure 1. Relationships between latex yield (T ha^{-1} yr^{-1}) and annual CO_2 sequestration (T CO_2 ha^{-1} yr^{-1})

Table 2 Annual CO_2 sequestration (-NEE), rubber latex yield, and estimated CO_2 sequestered per kg of latex from 2013 to 2016.

Year	Annual CO_2 sequestration (T CO_2 ha^{-1} yr^{-1})	Rubber latex yield (T ha^{-1} yr^{-1})	CO_2 sequestered per kg of latex (kg kg^{-1})
2013	28.0	1.20	23.3
2014	40.6	1.89	21.5
2015	43.1	1.71	25.2
2016	35.1	1.09	32.2
Mean	36.7	1.47	24.9

4. Conclusion

Annual CO_2 sequestration by rubber plantation ranged from 28.0 to 43.1 tons CO_2 ha^{-1} yr^{-1} and averaged 36.7 tons CO_2 ha^{-1} yr^{-1}. Thus, it is estimated that 2.95 million hectares of latex-producing natural rubber plantations in Thailand has potential to sequester approximately 108 million tons of CO_2 per year. On average, approximately 24.9 kg of net CO_2 was sequestered for each kilogram of natural-rubber latex produced. This large carbon sequestration potential associated with natural latex production has important implications for the establishment of 'environmental friendliness' marketing strategy to enhance the natural rubber competitiveness, especially in comparison to synthetic rubber.

5. References

[1] B Law *et al,* Agric For Meteorol 113, 113:97-120 (2002).
[2] DD Baldocchi *et al*, Bull Am Meteorol Soc 82, 2415-2434 (2001).
[3] DD Baldocchi, Global Change Biol 9, 479-492 (2003).
[4] DF Hui *et al*, Tree Physiol 23, 433–442 (2003).
[5] G.M. Lovett *et al.*, Ecosystems 9: 1–4 (2006).
[6] H Viriya, Asia-Pacific Journal of Science and Technology 22, 1-8 (2017).
[7] JC YANG *et al*, Acta Phytoecologica Sinica 29, 296-303 (2005).
[8] K Charoenjit *et al*, Procedia Earth Planet 7, 139-142 (2013).
[9] M. Aubinet *et al.*, Adv Ecol Res 30, 113–176 (2000).
[10] R Valentini *et al*, Nature 404, 861-865 (2000).
[11] SK DEY, The Indian Forester 131, 1429-1436 (2005).
[12] SR Maggiotto *et al*, Acta Sci Agron 36(2), 239-245 (2014).
[13] Z GUO *et al*, Agroforest Syst 66, 117-127 (2006).

Acknowledgements
This research was funded by the National Research Council of Thailand and the Thailand Research Fund, CIRAD and SOERE F-ORE-T (France). The authors thank colleagues from Chachoengsao Rubber Research Center, Rubber Authority of Thailand, French Agricultural Research Center for International Development (CIRAD), and Kasetsart University for their assistance.

Peptidomic and Bioinformatic Analysis to Evaluate the Application of Peptides from Latex of *Hevea brasiliensis* Clone BPM24

P Havanapan[1,*], N Phungthanom[1]

[1]Institute of Molecular Biosciences, Mahidol University, Salaya, Nakhonpathom, Thailand.

*Corresponding author: phattaraorn.hav@mahidol.ac.th

Abstract. Peptidomic analysis coupled with mass spectrometry identification was employed to investigate the constitutively expressed peptides in latex serum *Phytophthora*-resistant rubber tree (*Hevea brasiliensis*), BPM24. The utilization of proteins and peptides as a strategy to study new methods can lead to a therapeutic application. However, a high-efficiency method to extract peptides from latex serum is not well established. After optimization of latex serum isolation from natural rubber latex, using of lysis solution for peptide extraction exhibited superior serum peptide recovery than using other chemical reagents. Ultrafiltration followed by solid-phase extraction was selected to achieve high extraction yield of low molecular weight proteins and peptides. Mass spectrometric and bioinformatics analysis characterized and predicted the identified protein and peptide contents. All latex serum peptide fractions from C_{18} reversed-phase chromatography possessed effectiveness of antifungal activity against three tested plant fungal including *A. brassicicola*, *C. lunata* and *M. grisea* with different MIC in the range 250-2,000 µg/ml from different peptide fraction. Moreover, Etotal (E1-E5 pooled together) from lysis solution extraction was found to able to inhibit three tested plant fungal at MIC 500 µg/ml. Moreover, *B. cereus* and *E. faecium* were the sensitive gram-positive bacteria with MIC in the range 250-4,000 µg/ml. However, all gram-negative bacteria were not affected by 4,000 µg/ml. Among these tested pathogens, plant fungal was quite sensitive to our peptide product. Etotal from lysis solution extraction will be the best candidate for using further in antifungal testing of other plant fungal both *in vitro* and *in vivo*. The further investigation and application will lead to possible use of latex serum as a great biotechnological resource to develop peptide - based product for the treatment of plant infection.

1. Introduction

The widely cultivated Para rubber tree, *Hevea brasiliensis*, clone BPM24 produces a high yield of latex, is highly tolerant to *Phytophthora* disease. Therefore, it might be possible clone BPM24 must be produced antimicrobial proteins or peptides to protect itself from any pathogens. The term "peptidomics" was defined as the quantitative and qualitative analysis of endogenous or native low molecular weight proteins in biological samples [1]. However, peptide extraction efficacy of plant serum, especially *H. brasiliensis*, for industry application remains a technological challenge for increased coverage of the serum peptidome. The major problem for small peptide preparation is that their concentration is comparatively lower than the protein concentration. Peptide may diffuse from their tissue of origin and be diluted in rubber latex. Therefore, the specificity and sensitivity of processing for enhance the abundant peptides is necessary to further study. The aim of this research was to conduct an evaluation of peptide extraction method from *H. brasiliensis* latex serum to maximize the total number of peptides detected by mass spectrometric analysis together with the screening of extracted proteins for antimicrobial activity. These obtained and further data would be promising valuable peptides for new antifungal agent development.

2. Materials and methods

2.1 Latex serum isolation

Fresh latex samples were collected from the Para rubber trees, *H. brasiliensis* clone BPM24 by tapping method with Jeh-Bong knife. Thirty rubber trees, planted in the same crop at Khao Cha Mao district, Rayong province were selected for sample collection. To avoid contaminants, the fresh latex was collected directly into plastic container with or without addition of the preservative (0.6% ammonia). The latex serum was obtained by centrifugation of latex sample at 4 °C using the optimized condition. Then, the latex serum (supernatant) was collected and used for peptide extraction process.

2.2 Serum peptide extraction and subfractionation

To extract peptides from latex serum, serum sample was mixed with three different reagents at the final concentration of 80% acetone, 80% methanol, and lysis solution (8 M Urea, 2 M Thiourea, 2% CHAPS, 0.4% Triton X-100, 50 mM DTT. The filtrate was passed through consecutively on three molecular weight cut-off membranes (30, 10, and 3 kDa) using an Amicon ultra-15 centrifugal filter (Millipore, Germany). The filtrate (after passed 3kDa MWCO) was collected, pooled together, and stored in aliquots at -30 °C until used for peptide subfractionation by C_{18} reversed-phase chromatography. The filtrate mixed with 0.1% trifluoroacetic acid was subsequently loaded onto Solid-phase extraction (SPE) cartridges, SEP-PAK C18 reversed-phase chromatography [Waters]. The 50 % acetonitrile/0.1 % TFA was used as elution buffer. The eluate from the 50 % acetonitrile/0.1 % TFA elution were divided into five fractions, designated as E1 to E5, collected depending on volume to volume or pooled each fraction together (Etotal). Each fraction was concentrated and dried until becoming the powder by Freeze dryer [Thermo Fisher Scientific].

2.3 Peptidomic and Bioinformatic Analysis

The dried peptides were resuspended and vigorously mixed with MALDI matrix solution. The target samples were subjected to analysis in a Ultraflex III TOF/TOF mass spectrometer (Bruker Daltonics, Germany). The machine was run in the linear positive mode and mass spectra in the range of 500-5,000 m/z were collected. MALDI spectra of each sample from the mass spectrometer were subjected to spectral processing including peak detection, smoothing, baseline subtraction, and recalibration using flexAnalysis 3.0 software. For MS/MS analysis, the dried peptides were resuspended in 10 µL of 0.1 % formic acid. NanoLC–ESI MS/MS analysis was performed using an Ultimate 3000 LC system (Dionex, USA) coupled to an ESI-Ion Trap MS (Bruker Daltonics, Germany). These peptide fragment masses were detected and further processed by DataAnalysis software (Bruker Daltonics, Germany) to generate mass lists in the form of Mascot generic files. The MS/MS data from LC-MS were submitted to database search using the Mascot software *via* Matrix Science to identify peptide sequences. Moreover, the identified peptides were then blasted against a database of antimicrobial peptides through the bioinformatics program (AMPer) resource and find similarities to any known antimicrobial and anticancer peptides.

2.4 Antimicrobial testing

The pathogen that used in this experiment were 4 fungi [*Candida albicans* (ATCC 90028), *Alternaria brassicicola* (BCC 42724), *Curvularia lunata* (BCC 15558), *Magnaporthe grisea* (BCC 10261)] and 6 bacteria [*Acinetobacter baumannii* (ATCC 19606), *Pseudomonas aeruginosa* (ATCC 15692), *Klebsiella pneumoniae* (ATCC 700603), *Escherichia coli* (ATCC 25922), *Enterococcus faecium* (ATCC 51559),

Bacillus cereus (ATCC 11778)]. Antimicrobial activity was tested by broth microdilution assay. Briefly, each cell suspension was adjusted and further diluted to appropriated concentration. The assay was performed in 384-well plate in triplicate. Each well was added with extracted peptides (or positive or negative control agents) and cell suspension. Plates were then incubated at suitable temperature and hour depending on each microbial strain. Bacterial and fungal growth was observed by OD measurement at different excitation and emission wavelengths. Percent of bacterial/fungal inhibition was calculated. The concentration of crude peptide fraction required for 80% growth inhibition (MIC_{80}) was reported.

3. Results and discussion

3.1 Peptidomics analysis of H. brasiliensis clone BPM24 serum

Lysis solution extracted a higher number (93 ± 3) of low molecular weight peptides achieved more than that of 80% acetone and 80% methanol. Peptide extraction from fresh rubber latex by lysis solution followed by ultrafiltration (MWCO 30, 10, and 3 kDa, respectively) and then peptide purification and concentration (SEP-PAK C18) was the most efficient method to obtain superior serum peptide recovery than by other chemical reagents. There were some 31 peptides from 8 identified proteins found. The function of each identified proteins were also discussed. Most of identified proteins played roles in ATP synthesis and hydrolysis. The expression of ATP-binding cassette family was found to be high abundance in all extraction method of latex serum from pathogenic fungi tolerant *H. brasiliensis* BPM24. The bioinformatics program (AMPer) were then used to predict putative antimicrobial and anticancer peptides from the mass spectrometric-detected peptides. The prediction of peptide analysis revealed 15 putative antimicrobial or anticancer peptides. Most of these putative peptides showed broader spectrum activities or more potent activity against both Gram-negative and positive bacteria and fungi.

3.2 Antimicrobial activity of latex serum crude peptides

The fractionated peptides eluted from SEP-PAK column were collected. Each elution were subjected to perform antimicrobial testing. The results showed that all latex serum peptide fractions from C_{18} reversed-phase chromatography possessed effectiveness of antifungal activity against three tested plant fungal including *A. brassicicola*, *C. lunata* and *M. grisea* with different MIC 250-2,000 µg/ml. Moreover, *B. cereus* and *E. faecium* were the sensitive gram-positive bacteria with MIC 250-4,000 µg/ml. However, all gram-negative bacteria were not affected by 4,000 µg/ml. Among these tested pathogens, plant fungal was quite sensitive with MIC_{70} value as little as 125 µg/ml. The further investigation and application will lead to possible use of latex serum as a great biotechnological resource to develop bioactive agents for the treatment of human and plant infection.

4. Conclusion

Peptidomic analysis of latex serum *H. brasiliensis* BPM24 was investigated. Peptide extraction from fresh rubber latex without any preservatives by lysis solution extraction and then ultrafiltration passed MWCO 3 kDa prior to SPE using SEP-PAK C18 is the efficient method to obtain superior serum peptide recovery than using other chemical reagents. Moreover, the antimicrobial activity of latex serum proteins could be observed. Most of peptide fractions were highly active against all plant-pathogenic fungi and less active against gram-positive bacteria. Our perspective investigation and application will lead to possible use of latex serum as a great biotechnological resource to develop bioactive agents or peptide-based product for the treatment of *Phytophthora*-infected *Hevea brasiliensis* because various *Phytophthora* species cause leaf fall and black stripe diseases leading to a yield loss in rubber plantations.

IOP Conf. Series: Materials Science and Engineering **526** (2019) 012037 doi:10.1088/1757-899X/526/1/012037

5. Reference

1. Schulte I, Tammen H,Selle H, Schulz KP. Peptides in body fluids and tissues as markers of disease. Expert Rev Mol Diagn. 2005;5:145-157.

Acknowledgements
This work was financially supported by the Thailand Research Fund (TRF) and National Research Council of Thailand (NRCT) to PH.

IOP Conf. Series: Materials Science and Engineering **526** (2019) 012038 doi:10.1088/1757-899X/526/1/012038

Properties of natural rubber (NR) and wood/NR composites as gamma shielding materials

K Ninyong[1], E Wimolmala[1], N Sombatsompop[1] and K Saenboonruang[2,*]

[1] Polymer Processing and Flow (P-PROF) Research Group, Division of Materials Technology, School of Energy, Environment and Materials, King Mongkut's University of Technology Thonburi (KMUTT), Bangkok 10140, Thailand
[2] Department of Applied Radiation and Isotopes, Faculty of Science, Kasetsart University, Bangkok 10900, Thailand , e-mail: kiadtisak.s@ku.th, fscikssa@ku.ac.th

Abstract. In this work, gamma shielding, cure characteristics, and mechanical properties of natural rubber (NR) and wood/NR composites was investigated with the addition of lead (Pb) powder for potential use as flexible shielding materials. The content of Pb powder in this work was varied from 0 to 50 part per hundred parts of rubber by weight (phr) in 10-phr increment. The results showed that the increases in Pb contents led to the improvement in gamma attenuation but had negligible effects on cure properties of the composites. The increases in Pb contents reduced overall tensile properties of the NR and wood/NR composites, while the addition of 20-phr wood particles in wood/NR composites increased the tensile strength, elongation at break, surface hardness, and dimensional stability. Hence, the overall properties investigated in this work suggested that the developed NR and wood/NR composites could be used to attenuate low-intensity gamma rays produced after former steps of shielding such as after thermal neutron or beta shielding.

1. Introduction

Gamma rays are relatively high energy and high penetrating electromagnetic waves with their energy higher than 100 keV. Gamma rays are presently used in various applications including gamma imaging [1], elemental analysis based on PIGE [2], and radiosurgery [3]. However, despite all the usefulness of gamma rays, exposure to high doses could fatally harm users and works. The risks include nausea, vomiting, fatigue, skin burns, and even death [4]. In order to reduce risks of too high exposure to gamma rays, the radiation safety principle called "As Low As Reasonably Achievable" or "ALARA", which includes sufficient management in working time and distance, and effective radiation shielding installation and uses, must be strictly followed in all radiation-related facilities. Particularly for shielding, its importance is specially emphasized in the case of works that need extended period of time or close proximity to gamma sources. In principle, elements with high atomic numbers and densities such as lead (Pb) and Pb-containing materials are usually used as gamma shields due to their high probability of interaction with gamma rays and high energy transfer between gamma rays and nuclei of the heavy elements. Examples of gamma shielding materials are Portland cement with the addition of granulated Pb [5] and PbO-SiO_2 glasses [6].

Natural rubber has great potential to be utilized in such applications as NR is known to have high flexibility, great resistance to abrasion, and also superb toughness. Examples of NR used in radiation shielding applications were boron oxide (B_2O_3)/NR composites as neutron shielding materials [7], and Pb/NR composites as gamma shielding materials [8]. In spite of the advances in radiation shielding development from NR, there are still rooms for further improvement, especially in terms of dimensional stability and surface hardness. One possible method to achieve these additional properties is to introduce wood particles into the composites. Since wood particles have high rigidity, they could improve dimensional stability and increase stiffness as well as surface hardness. The wood/NR composites were previously used in roofing applications that showed significant improvements in strength, UV protection, and thermal insulation [9], hence, applying wood/NR composites with Pb powders could present new developed materials that are able to utilize in wider radiation shielding applications.

As a results, this work aimed to investigate gamma shielding, cure characteristics, and mechanical properties of NR and wood/NR composites with the addition of Pb powder, in which its contents were varied from 0 to 50 parts per hundred parts of rubber by weight (phr) in 10-phr increments. It should be noted that this work focused on the attenuation of gamma rays caused after the former step of neutron shielding (B_2O_3/NR composites), hence, the intensity of the incoming gamma rays was not as high as the ones from gamma sources.

2. Experimental

2.1 Materials and chemicals

Natural rubber (STR 20) was used as the main matrix for this work. The vulcanizing formulations, which include their roles, are given in Table 1. It should be noted that types of wood particles should be negligible effects on gamma shielding properties as they mostly contain light elements, however, have potential to improve strength of the materials.

Table 1. Material formulations of NR and wood/NR composites including chemicals and their roles.

Ingredient	Content (phr*)	Function
Natural rubber (STR 20)	100 (part)	Main matrix
ZnO	5.0	Activator
Stearic acid	2.0	Activator
Mercaptobenzothiazole; MBT	0.5	Accelerator
Diphenylguanidine; DPG	2.0	Accelerator
Carbon black (HAF N330)	40.0	Reinforcement
Wood particles	20.0	Filler
Silane coupling	0.5%wt of wood particles	Coupling agent
UV stabilizer (CHIGUARD 234)	0.1	UV absorber
Anti-oxidant (TY 1076)	0.5	Anti-oxidant
Anti-ozonant (REDEZON 515P)	3.0	Anti-ozonant
Sulphur	3.0	Cross-linker
Lead powder (Pb)	Varying from 0, 10, 20, 30, 40, and 50	Gamma protective filler

** phr: parts per hundred parts of rubber by weight*

2.2 Preparation of NR composites

The NR was masticated on a laboratory two-roll mill (Yong Fong Machinery Co., Ltd., Thailand) for 5 min. The masticated NR was then compounded with the chemicals given in Table 1. The compounding process was also carried out on the two-roll mill for 25 min to ensure the uniform distribution of all fillers. For wood/NR composites, 20-phr of wood particles was chemically treated with a silane coupling agent and oven-dried at 100°C for 6 h until constant weight was achieved before being mixed with NR [10].

2.3 Cure characteristics

The scorch time (t_{s1}), cure time at 90% (t_{c90}), and torque differences of the NR and wood/NR compounds were evaluated using an oscillating die rheometer (ODR GT 70-70-S2, GOTECH Testing machine, Inc., Taiwan) following ASTM D2048-11 standard testing at the temperature of 160°C. The determined cure times were then for vulcanizing the NR composites sheets in hydraulic press at 160 kg/cm^2 for preparation on NR composite specimen sheet.

2.4 Mechanical and physical properties

The tensile properties including tensile modulus, tensile strength, and elongation at break of the NR and wood/NR composites were investigated using a universal testing machine (Autograph AG-I 5kN, Japan) following ASTM D412-06 standard testing. For hardness measurement, all the samples were tested using a hardness durometer, Shore A (Teclock GS-719G, Japan) according to ASTM D2240-05.

2.5 Gamma ray shielding properties

The gamma ray shielding properties were performed at the Thailand Institute of Nuclear Technology (TINT), in which the incoming gamma rays were produced after thermal neutron shielding with 300-mCi ^{241}Am/Be as a neutron source and 80-phr B_2O_3/NR as neutron shielding materials. The gamma detector used in this work was a survey meter (LUDLUM 2200) connected to a set of high voltage power supply, amplifier, and timer/counter.

3. Results and Discussion

3.1 Cure characteristics

Cure characteristics of NR and wood/NR composites are shown in Table 2. The results indicated that adding Pb powder to the composites did not significantly change cure characteristics due to the fact that Pb does not chemically interact with NR matrix. Furthermore, it was observed that wood/NR composites had higher scorch time and cure time compared with NR composites at all Pb contents. This was because wood particles had hydroxyl groups that could absorb activators and accelerators, slowing the chemical vulcanizing process [7]. In terms of the torque differences, the results indicated that increasing Pb contents led to the increases in the values due to the high rigidity of Pb powder.

Table 2. Scorch time, cure time, and torque differences of NR and wood/NR composites with additions of Pb powder. *(Numbers after ± represent the standard deviation)*

Pb content (phr)	NR composites			Wood/NR composites		
	Scorch time (s)	Cure time (s)	Torque difference (dN m)	Scorch time (s)	Cure time (s)	Torque difference (dN m)
0	34±2	184±1	74±1	42±8	222±12	79±1
10	35±2	200±2	93±0	40±5	250±13	79±1
20	34±3	205±2	95±0	38±1	220±2	85±0
30	35±2	206±3	96±0	38±3	219±1	86±1
40	34±1	204±2	97±0	37±1	220±0	87±1
50	34±1	208±4	97±0	37±3	222±2	86±0

3.2 Mechanical properties

The results of mechanical properties are shown in Table 3, which suggested that increases in Pb contents reduced overall tensile properties of both NR and wood/NR composites, while the addition of 20-phr wood particles in wood/NR composites increased the tensile strength, elongation at break, surface hardness, and dimensional stability. The reduction in tensile properties caused by adding Pb powder were due to the poor interfacial compatibility of fillers.

Table 3. Mechanical properties of NR and wood/NR composites with additions of Pb powder. *(Numbers after ± represent the standard deviation.)*

Pb content (phr)	NR composites				Wood/NR composites			
	Tensile modulus (MPa)	Tensile strength (MPa)	Elongation at break (%)	Hardness (Shore A)	Tensile modulus (MPa)	Tensile strength (MPa)	Elongation at break (%)	Hardness (Shore A)
0	3.7±0.0	14.4±0.9	316±19	74±1	3.7±0.0	15.4±0.3	460±10	76±1
10	3.6±0.0	11.0±1.2	262±26	70±2	4.0±0.0	14.0±1.0	391±24	75±1
20	3.6±0.0	9.2±2.4	228±52	70±1	3.7±0.0	13.3±0.5	411±11	75±1
30	3.6±0.0	7.8±0.9	197±22	71±1	3.6±0.1	12.1±0.5	387±29	75±1
40	3.5±0.0	7.8±1.4	213±34	70±2	3.4±0.1	11.0±0.6	372±8	75±1

Pb content (phr)	NR composites				Wood/NR composites			
	Tensile modulus (MPa)	Tensile strength (MPa)	Elongation at break (%)	Hardness (Shore A)	Tensile modulus (MPa)	Tensile strength (MPa)	Elongation at break (%)	Hardness (Shore A)
50	3.5±0.1	6.4±1.3	178±33	71±1	3.5±0.0	10.9±0.2	368±8	76±1

3.3 Gamma shielding properties

Figure 1. Shown the gamma shielding properties of NR and wood/NR composites, Fig.1a indicates the values of linear attenuation coefficient (μ) of samples with different Pb contents while Fig.1b indicates the gamma transmission of samples with 50-phr Pb content at different thicknesses. The overall results implied that increasing Pb contents resulted in the improvement of shielding abilities as seen by the increases in the μ values. Furthermore, it was observed that the NR composites had slightly higher μ than the wood/NR composites due to the dilution effect of wood particles in wood/NR composites that reduced the capability of Pb powders to attenuate gamma ray. However, after considering gamma transmission shown in Fig.1b, negligible differences in gamma attenuation were observed at thicker sheets of samples, indicating similar gamma shielding properties between the two composites and great potential for wood/NR composites to be used as flexible gamma shielding materials.

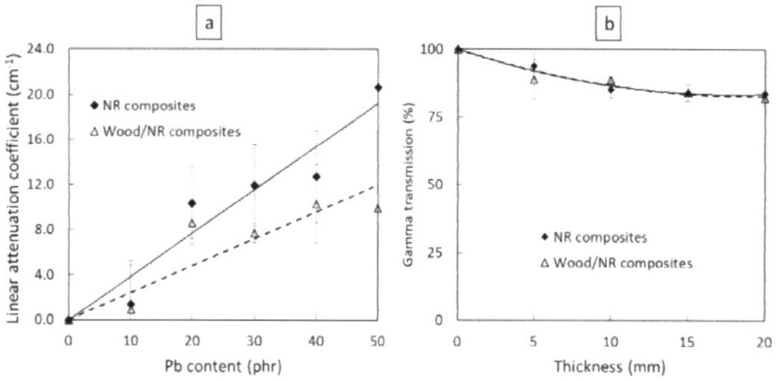

Figure 1.
(a) Linear attenuation coefficients for different contents of NR and wood/NR composites
(b) gamma transmission for different thicknesses for NR and wood/NR composites (50 phr Pb content)

4. Conclusions

Flexible gamma shielding materials based on NR and wood/NR composites with addition of Pb powder were developed. The results showed that wood particles improved overall mechanical properties and dimensional stability of the composites, but negligibly affected gamma shielding properties. Furthermore, the Pb powders effectively acted as gamma protective fillers as seen by the improvement in gamma attenuation with increasing Pb content. Based on the experimental results in this work, it could be concluded that NR and wood/NR composites had great potential to be used as flexible and effective gamma shielding materials.

5. References

[1] Zhu Z, Zheng Y, Peng Z and Zhang Y 2004 *Polym. Test.* **23** 217-23
[2] Chhillar S, Acharya R, Sodave S and Pujari P K 2014 *Anal. Chem.* **86** 11167-73
[3] Kruyt I J, Verheul J B, Hassens P E J and Kunst H P M 2018 *J. Neutrosurg.* **128** 49-59
[4] Walsh L 2013 *Radiat. Environ. Biophys.* **52** 29-36
[5] Hosiny F I, El-Faramawy N A 2000 *Radiat. Meas.* **32** 93-9
[6] Singh K J, Singh N, Kaundal R S and Singh K 2008 *Nucl. Instrum. Meth. Phys. B.* **266** 944-8
[7] Ninyong K, Wimolmala E, Sombatsompop N and Saenboonruang K 2017 *Polym. Test.* **59** 336-43
[8] Gwaily S E, Madani M and Hassan S S 2002 *Polym. Compos.* **23** 495-9
[9] Yamsaengsung W and Sombatsompop N 2008 *J. Macromol. Sci. Phys.* **47** 967-85
[10] Sombatsompop N and Chaochanchaikul K 2005 *J. Appl. Polym. Sci.* **96** 213-21

Acknowledgements
Authors would like to acknowledge financial supports from the Thailand Research Fund under Grant Number RDG5750096 and Grant number RDG5850094; and the Office of the Higher Education Commission (OHEC) under the National Research University (NRU) program. Authors also would like to thank the Thailand Institute of Nuclear Technology (TINT) for the technical support.

Impact strength enhanced bioplastic polylactic acid by interfacial bonding with hydroxylated natural rubber

P Boochathum[1], W Mingwanish[1] and S Pivsa-Art[2]

[1] Chemistry Department, Faculty of Science, King Mongkut's University of Technology Thonburi, Bangmod, Tungkru, Bangkok 10220, Thailand.
[2] Chemical and Materials Engineering, Rajamangala University of Technology Thanyaburi.

*Corresponding author: ploenpit.boo@kmutt.ac.th

Abstract. The effect of functionality of HNR and weight composition of PLA:HNR on the mechanical properties, impact strength, water vapour and oxygen gas transmission rates was studied. From morphology result, co-continuous phases in PLA:HNR050 and PLA:HNR035 were more obvious than PLA:HNR025. The observed co-continuous phases might be caused by the interfacial bonding between PLA phase and HNR phase. Better compatibility between HNR035 and PLA was believed to result in the considerably improvement of impact strength and slight reduction of mechanical properties of PLA. It is interesting that water vapour and oxygen gas transmission rates of PLA significantly increased with HNR035 loading. As a result, the appropriate weight composition of HNR035 blended with PLA was clarified to be 70:30 that was about tipple higher than the composition of NR as reported in the previous studies.

1. Introduction

Toughness properties of PLA could be developed by blending with epoxidized natural rubber (ENR) compared to NR[1]. The rubber phase of NR was clarified to disperse in the continuous PLA matrix with small droplet. Increasing of NR content, the large droplet size of rubber and the partially compatible between PLA and ENR were observed. The amount of NR at 10 %w was claimed to give optimum property. Wahit M U et al [2] prepared PLA/ENR-50 blends by melt extrusion followed by injection molding to fabricate the test samples. The toughness of the blend was improved with ENR-50 loading up to 20 %w but flexural and tensile strength decreased. SEM showed good distribution and increased ENR particle size as ENR content increased from 10 to 30 %w. Water absorption was observed to increase with ENR loadings. Recently, the synthesized NR-graft-PLA (NR–PLA) via esterification of maleated NR (NR-MAH) with PLA was used as a compatibilizer in the PLA/NR blends[3]. The addition of 5 %w NR–PLA [36.6 %w grafted PLA content] into PLA/NR blend [80/20 (w/w)] was found to increase Izod impact strength of PLA from 28.9 J/m to 62.7 J/m due to partial miscibility of blends.

This research aims to investigate the effect of the functionality of hydroxylated natural rubber (HNR) and the blend ratio on the compatibility and mechanical properties including impact strength of PLA. The water vapor and oxygen transmission rates of the PLA:HNR blends were also investigated.

2. Experimental

2.1. Materials

Polylactic acid (Grade 4043D, NatureWorks®). Natural rubber (low ammonia (LA) 60% DRC purchased from Num Rubber & Latex Co., Ltd., Thailand). Hydrogen peroxide (Analytical Grade 35% purchased from T.S.Interlab, Thailand). Formic acid (Analytical Grade 85% from Fisher Chemical, UK). Triton-x 100 (Fisher Chemical, UK). Liquid nitrogen purchased from Linde Co., Ltd., Thailand.

2.2. Preparation hydroxylated natural rubber

NR latex 500 g was added with titron-x100 50 mL followed by addition of the mixture of hydrogen peroxide (H2O2) and 85% w/w formic acid (CH2O2) using the mole ratio of rubber : hydrogen peroxide : formic acid = 1 : 0.25 : 0.25 for HNR025, 1 : 0.35 : 0.35 for HNR035 and 1 : 0.50 : 0.50 for HNR050. The reaction was carried out for 2 hours at 70oC, subsequently, the obtained HNRs were washed with distilled water until the pH of washed water became 7 and then dried at the temperature of 50-55°C.

2.3. Mixing PLA with HNR

Various compositions of PLA:HNR blends from 90:10, 80:20 to 70:30 were prepared by melt blending using the internal mixer at 190oC with a rotor speed of 40 rpm for 10 minutes. Subsequently, they were compression molded into sheets for further property testing under the condition with temperature 200°C, pressure150 kg/cm2 for 10 minutes.

2.4. Morphology study

Morphology of the blends was investigated using Scanning Electron Microscope (SEM) (JEOL, JSM-6610LV). Fresh surfaces were created by immersing the samples in liquid nitrogen and then the fractured samples obtained were coated with gold by sputter coater (Cressington, 108 Auto).

2.5 Mechanical property testing

The impact strength of PLA and their blends were tested by notched Izod impact tester according to ASTM D256. The tensile strength, elongation at break and modulus of them were also tested by universal testing machine (model LR 50K, LLOYD International) with a load cell of 50 kN and crosshead speed of 5 mm/minute according to ASTM D638.

2.6 Glass transition temperature determination

Glass transition temperatures were determined using Mettler-Toledo Differential Scanning Calorimeter (DSC) (Perkin Elmer Pyris-7). Heating was run from -100 °C to 300 °C with the rate of 10°C/minute.

3. Results and Discussion

The FTIR spectra clarified the presence of the -OH in the range of 3388-3409cm^{-1} and the C=O of the ester goups in the range of 1732-1735 cm^{-1} for HNR whereas these peaks were not found to present on the NR molecules[4]. The functionality of HNR were confirmed by the glass tansition temperatures as summarized in table 1 that increased with the number of the functional groups.

Table 1. Glass transition temperatures of HNR and NR.

Rubber type	T_g (°C)
HNR050	-46.22
HNR035	-57.98
HNR025	-59.24
NR	-63.94

Figure 1 shows clear droplets of HNR025 phase that distributed in the PLA matrix (c) whereas co-continuous phases in the molded PLA:HNR050 (a) and PLA:HNR035 (b) were obvious. However, the increasing HNR035 composition ratio was found not to change the morphology of the PLA:HNR035 blends as shown in figure 2. The co-continuous phases observed might be resulted from the interfacial bonding between PLA phase and HNR phase.

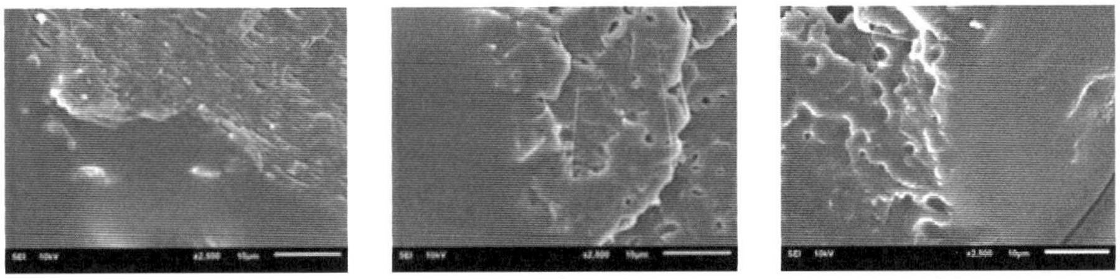

(a) (b) (c)

Figure 1. SEM images of PLA:HNR050 (a), PLA:HNR035(b) and PLA:HNR025(c) with the composition ratio of 90:10

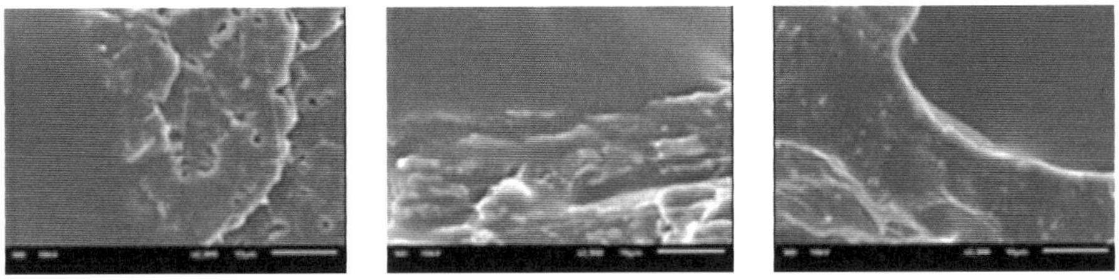

(a) (b) (c)

Figure 2. SEM images of PLA:HNR035 = 90:10 (a), 80:20 (b) and 70:30 (c)

The mechanical and impact strength properties of PLA/HNR compared to PLA were studied. Figure 3 shows that PLA:HNR035 gave the highest tensile strength (34.26 MPa), elongation at break (11.56%) and modulus (313.63 MPa) compared to those of PLA:HNR050 and PLA:HNR025. The appropriate ratio of PLA:HNR035 was found being 70:30 w/w that gave the highest tensile strength of 36.18 MPa, and the elongation at break of 23.47% as shown in the figure 4. On the other hand, it was significant that the impact strength of PLA was drastically improved by HNR035 loading as shown in figure 5. However, the impact strength of PLA:HNR035 was found to slightly decrease with HNR035 loading but still remain higher than that of PLA as shown in figure 6.

Water vapor and oxygen transmission rates of PLA:HNR035 and neat PLA films as summarized in table 2 were measured using the Moduluated Infrared Sensor at the condition regarding to ASTM F1249-06 and ASTM D 3985-05, respectively. It was remarkable that water vapor and oxygen gas transmission rates of both PLA:HNR035= 70:30 and PLA:HNR035= 60:40 significantly higher than those of neat PLA. As a result, HNR035 considerably enhanced the gas barrier property of PLA due to well incorporation of HNR in the PLA matrix.

Table 2. Water vapor and oxygen gas transmission rates

Samples	Water vapor transmission rate* ($g/m^2/day$)	Oxygen gas transmission rate** ($cm^3/m^2/day$)
PLA:HNR035 60:40	13	130
PLA:HNR035 70:30	11	22
Neat PLA[5]	116	511

*38°C, relative moisture 90% **23°C, relative moisture 0%

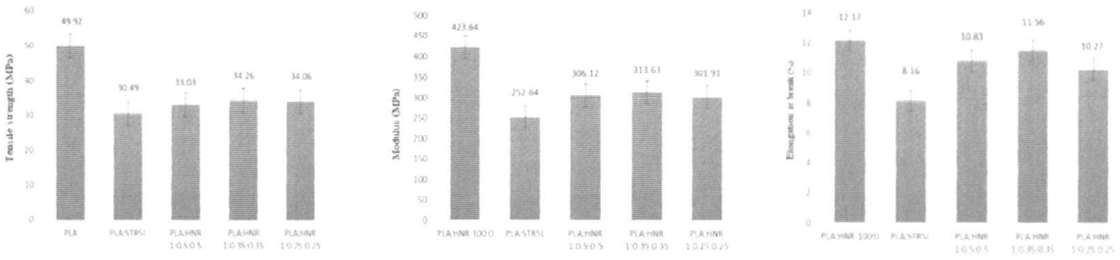

(a) (b) (c)

Figure 3. Mechanical properties of neat PLA and its blends with various types of HNR in the same composition of PLA:HNR=90:10.

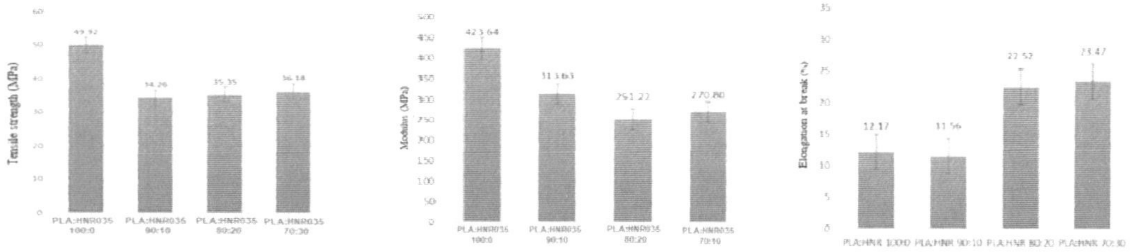

(a) (b) (c)

Figure 4. Mechanical properties of neat PLA and its blends with HNR035 at different compositions.

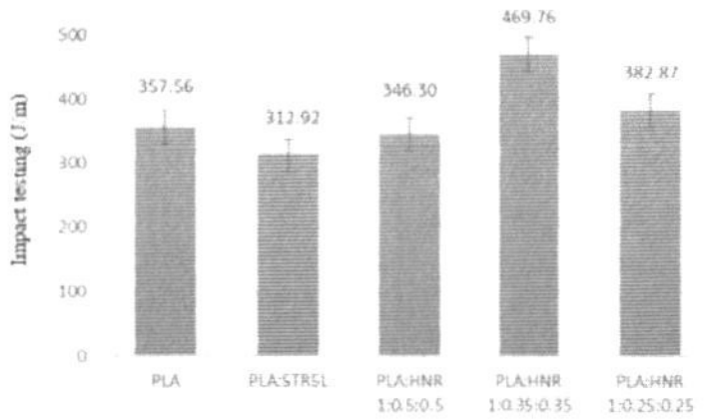

Figure 5. Impact strength of the PLA and its blends with various types of HNR.

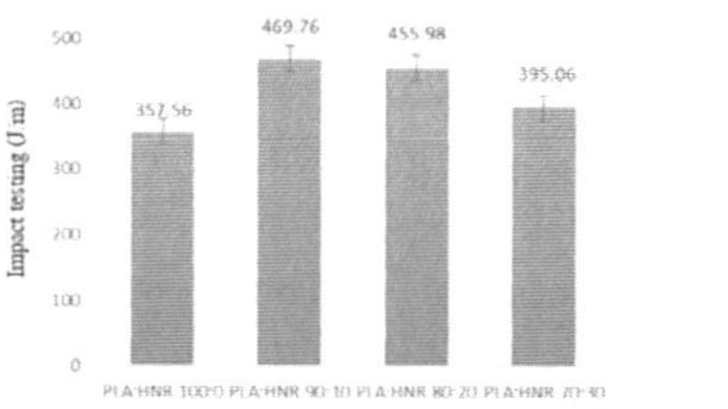

Figure 6. Impact strength of PLA and its blends with HNR035 at different compositions.

4. Conclusion

The observed co-continuous phases and the compatibility between PLA phase and HNR phase might be caused by the interfacial bonding. The remarkable improvement of impact strength and slight reduction of mechanical properties of PLA were obvious. HNR035 enhanced the highest impact strength of PLA with the appropriate blend ratio of PLA:HNR035 =70:30. Furthermore, HNR035 was also found to significantly improve the gas barrier property of PLA.

5. References

[1] Klankhetkon P, Chanchai T and Onuma S 2013 *Energy Procedia* **34** 888-897
[2] Mat U W, Azman H, Akos N I, Nurhayati A Z and Kayathre K 2015 *Sains Malaysiana* **44(11)** 1615–1623
[3] Phijittra S and Napida H 2017 *J of Materials Research* **32(4)** 788-800
[4] Ploenpit B and Pajaree C 2017 *The 13th Ecomaterials* Bangkok Thailand **Nov. 19-23** 93-99
[5] Chuenkhwan T and Somjai 2016 *Key Engineering Materials* **718** 10-14

Acknowledgement
Authors wishing to acknowledge the Thailand Research Fund for the financial support RDG5950021 for this research work.

Formulating policy recommendations for potential rubber plantation owners and rubber wood consumers in Thailand to obtain international forest management certification

K Duangsathaporn[1], P Prasomsin[1], N Laemsak[2], Y Omule[1], K Palakit[1], P Lumyai[1]

[1]Department of Forest Management, Faculty of Forestry, Kasetsart University, Thailand, e-mail: fforkcd@ku.ac.th
[2]Department of Forest Product, Faculty of Forestry, Kasetsart University, Thailand.

Abstract. The severity of environmental problems affecting air, soil, water, biodiversity, and forests, prompted relevant organizations to create mechanisms to prevent and manage these problems. One of these key mechanisms is Forest Certification, which is classified into two types, namely, Forest Management Certification (FM) and Chain of Custody Certification (CoC). The FM and CoC are certified by the relevant organization such as the Forest Stewardship Council (FSC) and the Programme for the Endorsement of Forest Certification (PEFC). These mechanisms will have an impact on the industry and commerce that use wood and tree products as a production base because of wood product importers may only import the certified products. This research project was designed to formulate policy recommendations for the potential development of rubber plantation owners and rubber wood consumers in Thailand to obtain international forest management certification. This was done using research data from two subprojects, namely, subproject 1: The analysis of rubber plantations on forest management certification status in Thailand, and subproject 2: Analysis of the factors affecting forest management certification of rubber plantations in Thailand. This resulted into four main policy recommendations in four broad areas: 1) Formulation of national policies and strategies to promote and access the forest certification systems for rubber plantation owners and rubber wood consumers; 2) Development of relevant organizations to promote and support forest certification in Thailand by determination and development of governmental organizations with the mission and potential in problem solving/negotiation and the potential developments of rubber plantation owners and rubber wood consumers for the whole system; 3) Development and availability of rubber plantation owners and rubber wood consumers in Thailand to obtain international certification; and 4) Research support in response to the development of forest certification in Thailand

1. Introduction

Due to the world's environmental problems including air, soil, water, and biodiversity, and the reduction of forest area problems, relevant organizations have created mechanisms to prevent and manage these problems. One of the key mechanisms is the Forest Certification, which is classified into two types, namely, Forest Management Certification (FM) and Chain of Custody Certification (CoC).

IOP Conf. Series: Materials Science and Engineering **526** (2019) 012040 doi:10.1088/1757-899X/526/1/012040

They are is certified by the relevant organization such as the Forest Stewardship Council [1] and the Programme for the Endorsement of Forest Certification [2]. These mechanisms will have an impact on the industry and commerce that use wood and tree products as a production base because of wood product importers may only import the certified products. The rubber tree (*Havea braziliensis*) is one of the most important economic plants in Thailand. Currently, Thailand is the world's largest producer and exporter of rubber products, and is the country's first revenue earner. Over the past decade, while rubber prices have declined, rubber plantations have continuously increased every year. In addition, rubber wood is currently exported to several countries, with most of the exports going to China. Thailand must solve the rubber price decline problem in both of wood and latex products. One important way is to increase the value of rubber wood products and open up marketing channels for exporting these products to other countries beyond China. Thailand is still unable to export rubber wood products to more foreign countries, such as Europe, USA and Japan, because most of the rubber plantations in Thailand are not certified for international forest management. Currently, there are only eight rubber plantations in Thailand, covering only 0.4% of the total plantation area, that have obtained Forest Management Certifications. This is very small when compared to other developed countries. This situation may affect the export rubber wood products because Thailand has less export channels and less power to negotiate which will increase the long-term risk of price and market competitions. In the U.S. Pacific Northwest, Tikina *et al.* [3] found three factors, namely, market pressure, land ownership pattern and water-body abundance, that influence the decision to obtain forest certification. While signalling and learning mechanisms lead to better and more transparent forest management and explain the movement towards certification, market incentives do not play an important role in the adoption of forest certification in Brazil [4]. It was differed from the study of Carlsen *et al.* [5] in Ghana who stated that customer demands, prospects of consolidating or increasing market shares, and a perceived need to invest in resource productivity were the main stated reasons for initiated certification uptake. In case of Thailand, it must accelerate to develop and promote the forest certification mechanism as a guideline to increase or expand the market for rubber wood product export and increase the bargaining power of products. However, one of the most important problems with the forest management certification in Thailand is that most of the rubber plantation owners and rubber wood consumers do not understand the FM principles, criteria and indicators enough until the FM auditor has made Corrective Action Requests (CARs) status or suspend a FM certification. Therefore, this research objective was designed to formulate policy recommendations for the potential development of rubber plantation owners and rubber wood consumers in Thailand to obtain international forest management certification using the information of forest management certification status and the factors affecting forest management certification of rubber plantations in Thailand.

2. Methods and Data Collection

The formulation of policy recommendations for the potential development of rubber plantation owners and rubber wood consumers in Thailand for obtaining international forest management certification was generated using the knowledge derived from 2 subprojects, namely, *subproject 1*: The analysis of rubber plantations on forest management certification status in Thailand, and *subproject 2:* Analysis of the factors affecting forest management certification of rubber plantations in Thailand. The subproject 1 revealed the analysis of the status of rubber plantations for forest management certification in Thailand. The primary and secondary data of rubber plantations in Thailand that included physical, biological, social, economic, and management issues were collected and were comparatively analysed along with the number and certified status with the classification of the Corrective Action Requests (CARs) in both of the major and minor CARs. Stakeholder analysis with their understanding and opinions on rubber forest management certification in Thailand was also investigated. The subproject 2, analysis of factors affecting forest management certification of rubber plantations in Thailand, consisted of stakeholder interviews and brainstorming in order to derive useful information on the factors influencing forest management certification analysis. All information from these subprojects formed an important database to formulate policy recommendations for the potential development of rubber plantation owners and rubber wood consumers in Thailand for obtaining international forest management certification. Before the final report was generated, the draft policy recommendations were subjected to public hearing meetings and this improved some contents following the comments of experts in related fields.

3. Results and Discussion

3.1 Status of rubber plantations for forest management certification in Thailand

From the data of forest management certification in Thailand (FSC, 2015; 2016; 2017; 2018), the number of rubber plantations that received the FSC-FM certification increased from 5 organizations in 2015 to 12 organizations in 2018, while the certified area of rubber plantations was only 0.396% (98,219 Rai) of the total area (24,824,746 Rai) (FSC, 2017). In 2017, the smallholders of rubber plantations (< 625 Rai) were grouped and passed the group certification for 2009 cases (0.155%) from the total of 1,294,995 cases, while the large holders of 86 cases participated and passed the certification for 8 cases. The causes of Corrective Action Requests (CARs) of rubber plantation for forest management certification in Thailand during 2011-2016 (FSC, 2016) were mostly found in Criteria 4.2, 6.4 and 6.7 for Major CARs and Criteria 1.1, 4.2, 8.3 and 8.5 for Minor CARs.

The Rubber Authority of Thailand (2016) divided stakeholders of rubber plantations for forest management certification in Thailand into six groups: 1) national, 2) socio-community, 3) farmers, 4) rubber entrepreneur, 5) trade partner and 6) staff. The analysis using the technique of stakeholder analysis divided stakeholders into four groups: 1) Government agencies, 2) NGOs, 3) farmers, and 4) rubber entrepreneur. The analysis of stakeholder interaction revealing the benefit or impact on forest management certification largely divided stakeholders into two groups:1) government and state enterprise agencies, and 2) people and private agencies. It was found that both of these two groups could receive benefits from forest management certification in rubber plantations. In case of the influence of several organizations on forest management certification of rubber plantation, the first group of government and state enterprise agencies influencing forest management certification included the Ministry of Natural Resources and Environment (MNRE), the Ministry of Agriculture and Cooperatives (MOAC), the Thai Industrial Standards Institute (TISI), the Royal Forest Department (RFD), the Rubber Authority of Thailand (RAOT), universities, and the Department of Agricultural Extension (DOAE). A second group of organizations influencing the forest management certification of rubber plantations of people and private agencies included the Forest Stewardship Council (FSC), the rubber business operator, certified organization, the Thai Hevea Wood Association, the Thai Rubber Association, the Thailand Forest Certification Council (TFCC), the Federation of Thai Industries (FTI), the Thai Furniture Association (TFA), the Rubber Network Council and Rubber Farmers Institute of Thailand (RNRF) and the Thai-Forest Certification Network (TCERN). Opinions and understanding of 734 rubber farmers and seven rubber business operators on forest management certification were divided to four groups. The first group of 322 rubber farmers who joined the forest management certification was not ready to achieve certified activities was 62.1%, while the management of the rotation period and re-planting is the common activity which they are ready to perform. The most problematic activities are the zoning of the conservation area, the understanding of the forest management practice based on the principle of forest certification and planning for wildlife resource survey in rubber plantations. The rubber farmers understood that the participation in the forest management certification shall be voluntary, area map shall be clear and long-term management plan shall be suitable and sustainable. They also believed that the forest certification did not increase operating costs and were delighted to join the forest certification if they have more rubber plantations in the future. The main problems of rubber farmers who joined the forest certification were plant diseases, prices, and numbers of certified holders. They need help from certified holders to distribute knowledge, give suggestions, increase the price and support rubber tree fertilizers.

The second group was rubber farmers who did not participate in the forest management certification. Those not ready to achieve certified activities was 98%, due to lack of forest certification knowledge, the complexity and condition of forest certification. However, they still have a good attitude towards forest management certification because the positive image of the certified plantations and products are more acceptable to both domestic and aboard customers. Although rubber farmers in this group understood less of the forest certification process, they knew the basic requirements of the forest certification such as the voluntary participation, sustainable forest management plan, and legal rights to operate the management unit with a clear map.

The third and the fourth groups were the rubber business operators with and without participation in forest management certification, respectively. The 4 rubber business operators were licensed the group certification of the FSC-FM with the operating cost of 300,000-400,000 Baht a year. The customers of these business operators were aboard buyers who demanded rubber wood products such as lumber, medium density fiberboards (MDF) and wood logs. Those business operators who had a positive attitude to the forest management certification was 75%, due to their acceptance from both domestic and international customers, the better forest management system and the image of the organization. Although some rubber business operators did not participate the forest management certification, they also clearly understood forest management certification and believed in the international certifications such as FSC and PEFC to increase incomes and welfares of employees. The main reason that the business operators in this group did not go into the system of the forest management certification was that customers did not request it.

3.2 Factors affecting forest management certification of rubber plantations in Thailand

Following the four groups of rubber farmers and business operators, the first group of rubber farmers participating in forest management certification was induced by the readiness in forest certification including the understanding the principle and benefit of forest certification (factor 1) and the increase in quality of life due to the increases of product sales and incomes (factor 2), although the forest certification was the cause of increasing unnecessary production costs (factor 3). The second group of rubber famers who did not participated forest management certification believed that forest certification system was not suitable (factor 1) but the forest management certification could enhance the image of the rubber forest management (factor 2). The main factors inducing the participation in forest management certification of rubber business operators were the corporate image promotion (factor 1), increased work effectiveness (factor 2) and increased income and product value (factor 3), respectively. In case of rubber business operators who did not participate the forest management certification, they believed that their organizations were not ready to participate the certification (factor 1), although they knew that forest management certification allowed products to be accepted by their customers (factor 2) and increase the corporate image as similar as the forest management system (factor 3).

3.3 Policy recommendations for the potential development of rubber plantation owners and rubber wood consumers in Thailand

Based on the information on section 3.1 and 3.2, the policy recommendations for the potential development of rubber plantation owners and rubber wood consumers to receive international forest management certification were developed as follows:

3.3.1 The formulation of national policies and strategies to promote and access to the forest certification system for rubber plantation owners and rubber wood consumers. The forest management standard for Thailand shall be developed and approved by the international organizations such as the Forest Stewardship Council (FSC). However, the adoption of FSC standard version 4 (Ver. 4) to version 5 (Ver. 5) without the formulation of the national standard will significantly affect the forest management certification in Thailand if there is no negotiation and preparation for change such as the acceptance for rubber planting together with other plants.

- The government shall promote access to the forest certification of rubber plantation owners and rubber wood consumers into the national strategies and reform plans. It shall be linked to the economic, social and forest resource development strategies.
- The government shall provide the appropriate policy revealing financial support and other manufacturing for the rubber plantation owners who entered the forest management certification system in cooperation with the private sectors operating rubber wood business.
- Local personnel shall be encouraged and developed to be the lead auditor of the international forest certification. It shall also be assigned an educational institution or other independent agency to act.

3.3.2 The development of the relevant organizations to promote and support forest certification in Thailand.

- Government organizations shall be defined and developed to facilitate problem solving/ negotiation and develop the potential of rubber owners and rubber wood consumers. The problem solving and negotiation tasks shall be carried out by the Thailand Rubber Authority as the main organization and integrated with the Ministry of Foreign Affairs, Ministry of Commerce and other related agencies. Mission to develop the capacity and availability of rubber plantation owners and rubber wood consumers in Thailand shall be conducted by the Thailand Rubber and other educational institutes.
- Certify Body (CB) in Thailand shall be developed since only two are currently available. This will have impact on the cost of certification and restrictions on the choice of rubber plantation owners and rubber wood consumers in Thailand to select.

3.3.3 The potential development and availability of rubber plantation owners and rubber wood consumers in Thailand to obtain the international certification.

- Provide appropriate guidance for the awareness and preparedness of rubber plantation owners and rubber wood consumers in Thailand to reach the international forest management certification association with the appropriate internal audit development. This task shall be conducted by the Thailand Rubber and other educational institutes.
- Established awareness and readiness to enter the forest management certification system of rubber plantation owners and rubber wood consumers in various levels. This task shall be implemented by the Thailand Rubber and other educational institutes.
- Publicize forest certification including certified holders to the public. This task shall also be implemented by the Thailand Rubber and other educational institutes.
- Promote the development of some specific equipment or tools in the country that are relevant to the rubber plantation management and meet requirements of the international forest certification. This task shall be conducted by the Thailand Rubber and other educational institutes.

3.3.4 The research support in response to the development of forest certification in Thailand

- The research shall develop the forest management certification standard for Thailand with the acceptance by the international organizations such as the Forest Stewardship Council (FSC).
- The research shall be consistent with and appropriate to the context of Thailand and the international certification system in both of principles, criteria and indicators.
- The research shall be focused on impacts of forest management in term of economic, social and environmental dimensions.

- The research shall be focused on impacts of rubber planting together with other crops and land use model of agroforestry systems in term of economic, social, and environmental dimensions.
- The research shall be related to the development of some specific equipment or tools in the country that are relevant to rubber plantation management and meet requirements of the international forest certification.

4. Conclusions and Recommendations

This research was designed to formulate policy recommendations for the potential development of rubber plantation owners and rubber wood consumers in Thailand obtaining international forest management certification using research data of the analysis of rubber plantations on forest management certification status in Thailand and the analysis of the factors affecting forest management certification of rubber plantations in Thailand. Four main policy recommendations were made. The first policy recommendation was the formulation of national policies and strategies to promote and access forest certification system for rubber plantation owners and rubber wood consumers (consisted of four approaches). The second policy recommendation was the development of relevant organizations to promote and support forest certification in Thailand through determination and development of governmental organizations for the mission and potential in problem solving/negotiation and the potential developments of rubber plantation owners and rubber wood consumers for the whole system (consisted of two approaches). The third policy recommendation was the potential development and availability of rubber plantation owners and rubber wood consumers in Thailand to obtain international certification (consisted of four approaches). The last policy recommendation was research support in response to the development of forest certification in Thailand (consisted of five research approaches).

5. References

[1] Forest Stewardship Council. 2016. FSC Principles and criteria. Available Source: https://ic.fsc.org/en/certification/principles-and-criteria, November 1, 2016.
[2] The Programme for the Endorsement of Forest Certification. 2010. Sustainable Forest Management (PEFC ST 1003:2010). Available Source: http://www.pefc.org/standards/, November 18, 2016.
[3] Tikina A, Kozak R and Larson B. 2008. What factors influence obtaining forest certification in the U.S. Pacific Northwest. Forest Policy and Economics 10: 240-247.
[4] Araujo M, Kant S and Couto L. 2009. Why Brazilian companies are certifying their forest. Forest Policy and Economics 11: 579–585.
[5] Carlsen K, Hansen C P and Lund J F. 2012. Factors affecting certification uptake – perspectives from the timber industry in Ghana. Forest Policy and Economics 25: 83–92.

Acknowledgements

This research was funded by the Thailand Research Fund (TRF). We also thank the staffs from the Laboratory of Tropical Dendrochronology (LTD), Faculty of Forestry, Kasetsart University for data preparation, collection and analysis.

IOP Conf. Series: Materials Science and Engineering **526** (2019) 012041 doi:10.1088/1757-899X/526/1/012041

Research to Develop International Standards for Rubber Household Glove and Rubber Thread

S Wirasate[1,*], O Pinprayoon[2], P Sunintaboon[1] and C Nimsuwan[3]

[1]Department of Chemistry and Rubber Technology Research Centre, Faculty of Science, Mahidol University, Bangkok 10400, Thailand.
[2]Department of Science Service, Ministry of Science and Technology, Bangkok 10400, Thailand.
[3]National Metal and Materials Technology Center, Pathum Thani 12120, Thailand.

* Corresponding author: supa.wir@mahidol.ac.th

Abstract. The international standards for rubber household glove and rubber thread were firstly initiated as New Work Item Proposal (NWIP) standard drafts. In order to obtain the ISO international standard, several levels of the standard drafts (NWIP, CD, DIS and FDIS) have to be achieved accordingly. To develop the standard drafts, the existing national standards ((TIS 2476-2552 for rubber household glove specification and TIS 2556-2554 for rubber thread specification) and information gathering from the meetings among the involved parties, e.g., producers, users and academic experts, visiting the companies, and product sampling test results were analyzed and standard drafts were prepared. For each level of the standard drafts, the drafts were submitted to the ISO representative for circulation to the authorized ISO members for comment and voting. Then, in the ISO meeting all comments were answered, rebutted and discussed. After the discussion in the ISO meeting, the conclusion related to the standard draft was given. The international standards for rubber household gloves (ISO 20057 Rubber household gloves – General requirements and test methods) and rubber thread (ISO 20058 General purpose rubber thread – Specification) have been published and used since 2017.

1. Introduction

Thailand has been the world's number one producer and exporter of natural rubber. Among rubber products in Thailand, nation's income from rubber latex products (i.e., glove and rubber thread) holds the 2nd position. Furthermore, rubber glove and rubber thread are the major products that domestically employ high amount of natural rubber. Thailand is the world's number one and number two exporter of rubber thread and gloves, respectively [1].

Development of the international standards for rubber products is important not only to protect consumers, but also to enhance the ability of producers to improve their product quality. International standards (ISO) related to specification of rubber household glove and rubber thread have not yet been available. Only national and regional standards are available. Thailand is one of the countries that has the specification standards of these two products (i. e., TIS 2476-2552 Rubber household gloves, TIS 2556-2554 for rubber thread specification) [2-3]. Therefore, the purpose of this work was to develop the international standards for rubber household glove and rubber thread.

2. Procedure and output

To develop the standard drafts, the existing national and regional standards including Thai Industrial Standards ((TIS 2476-2552 for rubber household glove specification and TIS 2556-2554 for rubber thread specification) were reviewed and information gathering from the meetings among the involved parties, e.g., producers, users and academic experts, company visiting, and product sampling test results were analyzed. Then, the New Work Item Proposal (NWIP) standard drafts were prepared. For obtaining the ISO standard, several levels of the standard drafts, namely NWIP, Committee Draft (CD), Draft International Standard (DIS) and Final Draft International Standard (FDIS), have to be achieved step by step. Figure 1 and Table 1 show the common steps and activities with outputs during the process to obtain the ISO standards, respectively.

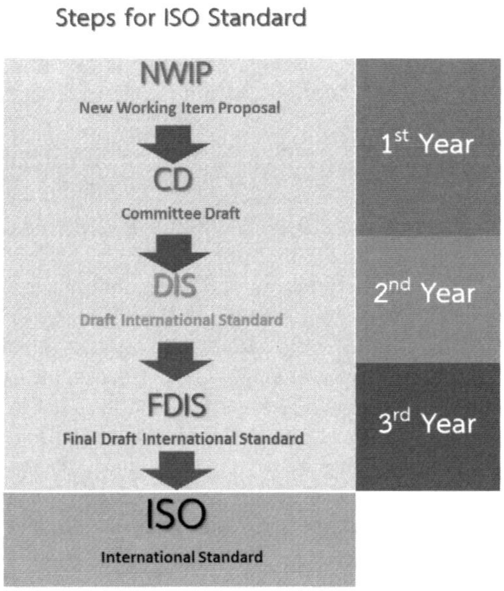

Figure 1. Steps to obtain the ISO standard.

Table 1. Activities and outputs during the process to obtain the ISO standards

Activities	Output
1. Existing standard review (national, regional and international standards)	- Only national and regional standards for rubber household glove were found. - Only national standard for rubber thread specification was found.
2. Meetings among the invlved parties, e.g., producers, users and academic experts and company visiting 3. Commercial product sampling tests	- Standard drafts (for each level) with answers to the comments from ISO members (for both activities 2&3).
4. ISO/TC (Technical committee) 45 meeting (once a year)	- Agreement on each standard draft and approval to go to the next step.

Both rubber household glove and rubber thread were classified based on their performance. The glove was classified into class 1A or 1B or 2A or 2B according to their initial and after ageing tensile properties (i. e., tensile strength and elongation at break) and tensile properties after immersion in general household liquids, including water, diluted surfactant and diluted acid solutions. For rubber thread, it was classified into 3 classes (class 1, 2 and 3) depending on initial tensile properties (i. e., modulus at 300% strain, tensile strength, and elongation at break) and after ageing tensile properties (i. e., tensile strength, and elongation at break) and initial tension set. For details of the specification of both rubber products, please see ISO 20057-2017 and ISO 20058-2017.

3. Conclusions

ISO 20057-2017 Rubber household gloves – General requirements and test methods [4] and ISO 20058-2017 General purpose rubber thread – Specification [5] have been published and used since 2017.

4. References

[1] Global Trade Atlas 2017.
[2] TIS 2476-2552 Rubber household gloves
[3] TIS 2556-2554 for rubber thread specification
[4] ISO 20057-2017 Rubber household gloves – General requirements and test methods
[5] ISO 20058-2017 General purpose rubber thread – Specification

Acknowledgements

Financial support from The Thailand Research Fund /the Commission on Higher Education (RDG5750014, RDG5850091 and RDG5950050) was gratefully acknowledged. Additionally, the authors would like to express our great appreciation to Thai Industrial Standards Institute for their supports throughout this work.

IOP Conf. Series: Materials Science and Engineering **526** (2019) 012042 doi:10.1088/1757-899X/526/1/012042

Innovative Design of Latex Gloves Production Lines via Modular Industrial-sized Prototypes

P Yongyingsakthavorn[1], T Chantrasmi, T Sriyubol, P Nimdum, A Tohsan and U Nontakaew

[1] Development of Machinery and Industrial Equipment Research Center (DMIE), and Faculty of Engineering, King Mongkut's University of Technology North Bangkok, 1518 Pracharat 1 Rd., Bangsue, Bangkok, 10800 THAILAND

*Corresponding author: pisit.y@eng.kmutnb.ac.th

Abstract. An innovative design of continuous latex gloves production lines was introduced in work. A modular approach to the designing process was described and key technologies were identified. Based on these technologies, industrial-sized prototypes were constructed and tested. The objectives were largely to reduce the production cost by some means such as reducing the thermal energy usage, shortening the length production line and time and decreasing the number of stationed workers. As a proof of concepts, prototypes of drying oven and stripping machine were constructed and the vulcanization experiments were carried out. The results were promising, for instances, the new drying oven with outside chain was projected to be able to reduce the thermal energy loss by approximately 12% compared to the traditional design and the stripping machine was shown to be able to vertically remove latex gloves from formers while occupying less space and potentially using fewer workers. From the vulcanization experiment, it was shown that the vulcanization time could be significantly shortened by pre-vulcanization and drying process.

1. Background and Introduction

Latex medical gloves industry has been growing continuously in Thailand, the largest manufacturer and exporter of raw-material rubber with one third of the total exports in the world. The demand for latex gloves worldwide rises an average of 5-10% per year or roughly 1,000 million pairs per year [1]. Meanwhile, it has been found that several Thai gloves manufacturers face many problems, such as higher production cost, both in term of materials and energy usage as well as from inefficiently old technologies.

To achieve competitiveness, our researchers and engineers are working to improve the production process starting from conceptual design, detailed design, manufacturing and finally testing. Many researchers were focusing on the chemical aspects of the process such as finding new latex formula and the effect of filling to reduce the material cost [2]. Little attention had been paid to the mechanical aspect of the production lines. In this work, the emphasis was on the mechanical aspect, particularly to find innovative solutions to improve each module of the powder-free production lines. As a constraint, the new solutions must be capable of producing at least 1 million pieces/day (42,000 pieces/hr).

The researchers have defined six key modules, namely stable conveying system, effective stripping and counting, latex dipping, former cleaning, drying and vulcanization. Currently, three out of the six – drying, stripping and vulcanization – were selected for detailed scrutiny and practical verification. In this phase, modular industrial-sized prototypes were to be fabricated as proofs of concept. The content below is divided into four more sections. The first three were organized according to the selected modules and the last section was the concluding remarks.

2. Drying and Vulcanization Ovens

One of the key modules to reduce the production cost by energy usage reduction is drying and vulcanization as discussed above. In the process of drying and vulcanization, a number of heated ovens are typically employed throughout the production line. These ovens operate in the temperature range of 90-130°C [3] and can be quite humid inside due to the trapped water vapor from the formers. Theoretically, the heat energy required – to raise the temperature of and evaporate the liquid water on the various stages of in-the-making latex gloves as well as on the former surfaces – is very small but there is usually a huge energy loss in the system.

Besides traditionally considered forms of heat loss through the insulated oven walls and convectively through hot air leakage, it has been identified that energy loss due to the conveyor chain plays a very significant role [4]. The conveyor chain – as well as additional pieces connecting the moving chain to the formers – is made of metal and thus can store a large amount of heat energy while it remains inside a heated oven. Once it emerges, it has a higher temperature than its surroundings (i.e. air at around room temperature) and therefore releases the stored heat energy back outside the oven, resulting in the energy loss. This is the so-called *convective* heat transfer by the conveyor chain in which the solid chain acts as a moving carrier of the thermal energy – as opposed to a *conductive* loss mode where heat is conducted along the length of the chain from inside to outside of the oven.

In this innovative design of the ovens, our researchers and engineers design an oven with outside conveyor chain as to avoid the above-mentioned heat loss. Figure 1 shows a traditional oven with inside conveyor chain. The innovative oven is divided into two parts with smaller drying volume so that the conveyor chain is arranged outside. Although it is clear by design that this will drastically reduce the convective heat loss, the new design also features more wall surface area and possibly more air leakage since, along the length of the conveyor chain, there is a long slot through which the connecting pieces between the chain and the formers move. As a result, heat loss through other modes might be increased. Therefore, at the conceptual level, it was not immediately clear whether the new design will result in energy saving in an actual practice. An industrial-sized prototype of this design is then constructed to collect data in order to answer this question.

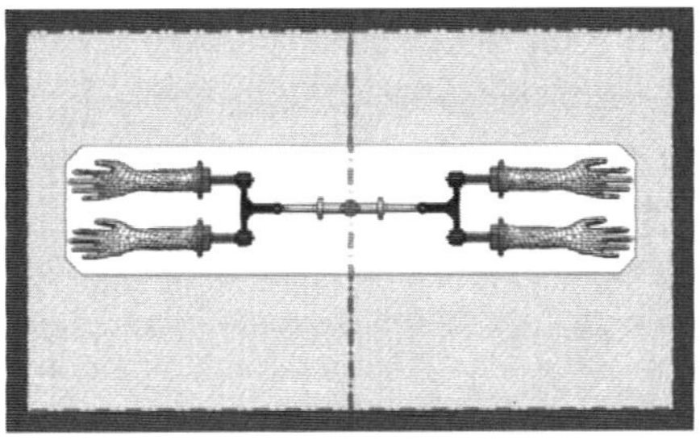

Figure 1. Schematic drawings of a traditional oven

The experimental setup for industrial-sized prototype of this new oven design is established, the prototype features a 20-meter conveyor chain in a simple loop through with 12 meters of chain inside the oven. It is a stand-alone prototype completed with hot-air delivery system, conveyor chain driving system and automatic control with a touch-screen panel. The prototype was then built accordingly and testing data were collected. The results were analyzed and showed that it was structurally sound, it was concluded that the new design would save approximately 12% of energy in an example production line [5]. This quantitative calculation assumed that the ovens of this type were to replace all the existing ovens in the production line. This would result an overall reduction of conveyor chain temperature everywhere (in other modules as well). A simple theoretical model was used to predict this new temperature profile and the quoted 12% saving was then based on this model [5]. Note that if this theoretical model was not used and the temperature of the chain was the same as with the traditional ovens, the energy saving would be well over-estimated.

3. Stripping Machine

Another identified key module essential to the innovative design is the stripping module. As the module's objectives, the new design must be (i) able to change its pitch distance to synchronize with the conveyor chain's pitch that changes slowly over time, (ii) able to reduce the required space both lengths wise an span wise, and (iii) easy to combine with the counting and packaging modules so as to reduce the number of stationed workers required, thus saving the production cost. The prototype is capable of vertically stripping formers from a double-former production line, meets the above objectives and is functionally equivalent to four conventional (horizontal) stripping machines commercially available in market.

4. Vulcanization

Generally, vulcanization of latex gloves consists of three main process, which are pre-vulcanization (maturation), drying and curing. In order to shorten the vulcanization time of latex gloves, the pre-vulcanization and drying processes were focused in the present work. First, an effect of pre-vulcanization time was evaluated by using swelling test as shown results in Figure 2 (a). It was observed that percentage of total solid content (% TSC) has no effect on pre-vulcanization time, i.e., equilibrium state was reached after 24 hours. In the case of filled sample, the pre-vulcanization time was reached after 36 hours. This result suggested that the presence of filler tends to prevent the pre-vulcanization of latex compound. Practically, in the production line, the pre-vulcanization is set at 60-80 hours. Therefore, from this experimental results, it may suggest that the pre-vulcanization time can be approximately decreased into 50% shorter.

In term of drying, evaporation flux of water in the latex film was calculated [6]. The obtained results are shown in Figure 2 (b). A similar tendency with swelling results was observed. It was found that % TSC has no effect on evaporation flux, i.e., the drying was completed after 5 min. In the case of filled latex film, the drying was extended to 7.5 min which is 50% shorter than the drying time that use in the production line.

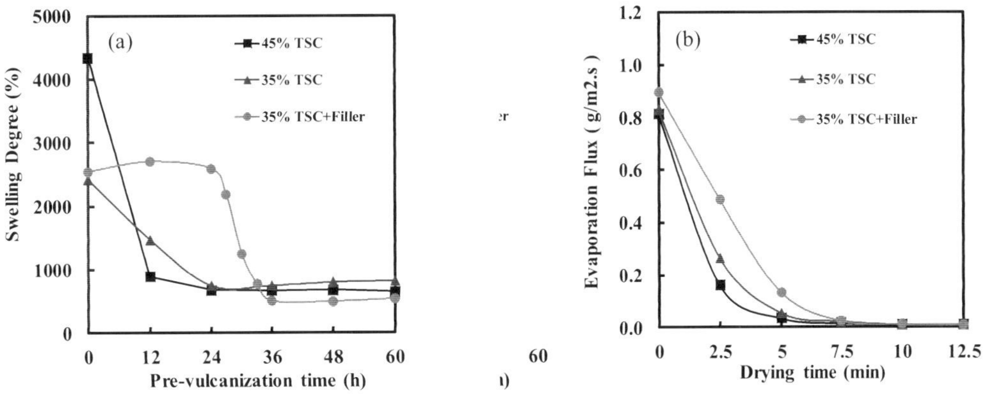

Figure 2. (a) swelling degree and pre-vulcanization time, and (b) evaporation flux and drying time.

5. Conclusions

A modular approach to the designing process was described and key technologies were identified. Industrial-sized prototypes of drying oven and stripping machine were constructed and tested. They showed that the production cost could be potentially reduced by employing these new designs. The new drying oven has the conveyor chain outside the heated oven area and was projected to reduce the total energy loss by approximately 12% on an example production line. The stripping machine could vertically strip latex gloves out of formers while demonstrably occupying less space and thus potentially using fewer stationed workers. The results from the vulcanization experiment showed that the required time could be significantly shortened by 50% via pre-vulcanization and drying process.

6. References

[1] Rubber Glove Newsletter, by THAI RUBBER GLOVE MANUFACTURERS ASSOCIATION, Vol. 38 1st Quarter, Jan-Mar 2014.
[2] Swinyard PE and Gorton ADT 1987. *NR Technology* **18** 41
[3] Kajornchaiyakul V 1989 *NR gloves manufacturer* (Songkhla Rubber Research Center: Songkhla)
[4] Final Report, "Conceptual Design of Production Line for Advanced Manufacturing of Natural Rubber Medical Glove", RDG5750133, supported by National Research Council of Thailand and The Thailand Research Fund.
[5] Final Report, "Industrial-Scale Prototypes of Ovens and Stripping Machines for Advanced Manufacturing Lines of Natural Rubber Medical Gloves (Phase 2)", RDG5950036, supported by National Research Council of Thailand and The Thailand Research Fund.
[6] Bouyer D, Philippe K, Wisunthorn S, Pochat-Bohatier C and Dupuy C 2009 *Drying Technology* **27** 59

Acknowledgement

The authors gratefully acknowledge the contribution of the National Research Council of Thailand and The Thailand Research Fund to this research project by providing financial support. Also, the authors would like to thank to the two rubber-glove factories that support various facilities to the project, and to assistant researchers/graduate students, W Koranuntachai, W Jitwiriya, M Chankrachang, and S Komonhirun, who mainly contribute to the research work.

IOP Conf. Series: Materials Science and Engineering **526** (2019) 012043 doi:10.1088/1757-899X/526/1/012043

A game-theoretic model to select suitable rubber policy

N Paoprasert[1], **T Varasirikul**[1], **J Pichitlamken**[1]
[1] Department of Industrial Engineering, Faculty of Engineering, Kasetsart University, Bangkok, Thailand
[2] Business Process Improvement, Feed Ingredient Trading Business Group, Charoen Pokphand Foods PCL, Bangkok, Thailand

*Corresponding author: fengnpp@ku.ac.th

Abstract. In this study, we used a game theoretic modelling technique to analyze the governmental policies by collecting natural rubber related data to create a model in order to select the policies that suit Thai rubber's stakeholders. After the model has been created, we used sensitivity analysis to capture the uncertainties of parameters such as the success probabilities of the policies and the rubber price. The results from the game theoretic model show that there are two policies that should be selected by the government which is the policies that support the increase of natural rubber consumption by 10%-15%. However, in order for the government to optimally select these policies, the success rate of the policies needs to be at least 30%. Otherwise, none of the policies should be selected. Regarding the policy to support Tier-1 entrepreneurs, since the increase in demand by implementing this kind of policy is fairly small, the policy should not be chosen. The policy for Tier-2 companies owned by the local farmers should be supported when the chance of selling the products is more than 35%.

1. Introduction

After the rise of the rubber price to its highest in 2011, the price has gone down ever since. There were strikes several times among the rubber farmers to ask for government's subsidy in the price. Providing subsidization may help absorbing farmers' expenses in a short run, but the government also explores some other potential solutions. The government also allocated several funding for several projects to support Thailand rubber farming situation. This includes soft loan projects for rubber processing business and the projects to help adding value to the natural rubber for business own by the local farmers.

To our knowledge, there has not been any research study focusing on applying a game-theoretic model to identify suitable governmental policy for rubber farming. In other countries, there have been numerous studies regarding to establishing game-theoretic models to study governmental policies. Most of the studies have been focusing on strategic decision in protection against terrorism ([1], [2], [3], [4] and [5]). The mentioned references are focusing on establishing a game-theoretic model for budget allocation against terrorism. Other studies such as Azaiez and Bier [6] use game theory to model resource allocation problem in terms of the reliability system. Brown et al. [7] study the optimal allocation on the US infrastructure.

IOP Conf. Series: Materials Science and Engineering **526** (2019) 012043 doi:10.1088/1757-899X/526/1/012043

2. The Proposed Model

There are 3 stakeholders in the game. The first one is the policy maker who needs to select the optimal rubber policies from the list of the available policies. The second player is the small and medium entrepreneurs in Thailand who process natural rubber to final products such as rubber gloves, condoms, rubber products from mould pressing, etc. We call this player the Tier-1. The third player is the companies who process natural rubber and supply the products to companies in the Tier-1. We call this player the Tier-2. The Tier-2 companies are mostly owned by the local farmer coops except for the latex company which is usually owned by a private investor.

The policy maker needs to make a decision about which policies to select (can be more than one) and how much to provide the budget for each product type under the budget constraint and the constraint that all farmers need to have the income that pass the poverty line.
The policy maker's objective function is shown below. In Equation (1), it is assumed that the estimated increase in the income (represented by the parameter p_{ijk}) cannot be fully achieved since it needs to be multiplied by the success probability of the policy and the portion of budget allocated.

$$\max \sum_{j\in\{T_1\cup T_2\}} \sum_{k=1}^{N_j} \sum_{i=1}^{10} (\gamma_{ijk} p_{ijk} \tau_{ijk} - b_{ijk}) x_i y_{ijk} \tag{1}$$

subject to the following constraints:

$$\sum_{j\in\{T_1\cup T_2\}} \sum_{k=1}^{N_j} \sum_{i=1}^{10} b_{ijk} x_i y_{ijk} \le Z \tag{2}$$

$$(f_l(r+hx_6)-((c_l-gx_5)a_l)\ge \mathrm{W}, \forall l \tag{3}$$

$$y_{ijk} \le x_i, \forall i, \forall j, \forall k \tag{4}$$

$$\sum_{k=1}^{N_j} \tau_{ijk} = 1, \forall i, \forall j \tag{5}$$

$$x_i \in \{0,1\}, \tau_{ijk} \ge 0, \forall i, \forall j, \forall k \tag{6}$$

γ_{ijk} = The success probability in achieving the goal estimated by the government from the implementation of project i to player type j who produces product type k
p_{ijk} = The estimated increase in income from the implementation of project i to player type j who produces product type k
τ_{ijk} = The decision variable regarding to the portion of budget to implement policy i to player type j who produces product type k
b_{ijk} = The cost of implementing project i to player type j who produces product type k
x_i = The decision variable of the policy maker to select project i
y_{ijk} = The decision variable of the type j^{th} player who produces product type k to follow the policy of project i
Z = The total budget
f_l = The amount of natural rubber per rai produced by farmer of size l
r = The rubber price (Baht per kg)
h = The government subsidy for natural rubber price per kg
c_l = The cost of natural rubber production per rai of farmer size l
g_l = The governmental subsidy for natural rubber production cost per rai
a_l = The average plantation area (in rais) of farmer size l
W = Thailand's poverty threshold
i = The policy index, i = 1, 2, ...,10
j = The index for the Tier-1 and Tier-2 players where $j \in T_1$ is for Tier-1 players and $j \in T_2$ is for Tier-2 players
k = The index for players who are processing product type k
l = The index for the farmers' sizes
N_j = The number of product types for Tier-1 or Tier-2 player.

During the time that the research study was conducted, there were ten available policies to select as shown in Table 1.

From Table 1, since policies 4 and 7 are too broad and vague, we exclude these two policies from our study (i.e., $x_4 = 0$ and $x_7 = 0$). Moreover these two policies require relatively small amount of budget, hence it does not affect the optimal solution. Also in Equation 3, farmers' sizes are assumed to be small, medium, and large.

The Tier-1 or Tier-2 player j has the following objective function. In Equation (8), it is assumed that the production capacity is increased by the portion of budget allocated to product type k.

$$\max \sum_{k=1}^{N_j} e_{ijk} \sum_{i=1}^{10} q_{ijk} d_{jk} \gamma_{ijk} \tau_{ijk} x_i y_{ijk} \quad (7)$$

subject to the following constraint

$$d_{jk} + \sum_{i=1}^{10} [q_{ijk} d_{jk} \gamma_{ijk}] \tau_{ijk} x_i y_{ijk} \le V_{jk} + \sum_{i=1}^{10} (V_{jk} \tau_{ijk}) x_i y_{ijk}, \forall j, \forall k \quad (8)$$

$$y_{ijk} \in \{0,1\} \quad (9)$$

e_{ijk} = An estimated increase in profit per unit after implementing policy i to player type j who produces product type k
q_{ijk} = The percentage of increase in demand after implementing policy i to player type j who produces product type k
d_{jk} = Initial demand of player type j who produces product type k
V_{jk} = Initial capacity of player type j who produces product type k

Table 1. The list of the available policies

No.	Policy explanation
1.	A project to increase the working capitals for Tier-2 companies who were processing natural rubber to latex
2.	A project to provide soft loans to increase capacity for Tier-1 companies who were processing natural rubber to final products such as tires, rubber gloves, and rubber bands
3.	A project to provide soft loan to increase capacity for Tier-1 companies own by the local coop who were processing natural rubber to Tier-2 products such as rubber sheets and rubber blocks.
4.	A project to support farmers to earn side incomes
5.	A project to subsidize partial cost of agricultural supplies (fixed amount per rai)
6.	A income subsidization project (fixed amount per rai)
7.	A project to support farmers to earn side incomes based on the sufficiency economy philosophy
8.	A project to revise the natural rubber market structure (with the aim to increase rubber demand by approximately 10%)
9.	A project to reduce 400,000 rais of rubber plantation area
10.	A project to setup the rubber process plant with the goal to increase rubber demand by 15%

3. Results and Discussion

The model is solved and sensitivity analysis is applied due to several uncertain parameters. These parameters are assumed to have ranges from the relatively estimated low values to the relatively estimated large values.

The rubber price is assumed to range from 10 Baht per kg to 150 Baht per kg. Other model parameters such as the plantation cost, the production capacity, and the plantation areas are also assessed. There are a total of 11,520 scenarios to run from the ranges in Table 2. The results are shown in Figure 1.

The figure above shows the percentage that each policy is chosen for each rubber price ranging from 10 Baht per kg to 150 Baht per kg. The result shows that policy 10 is chosen the most followed by policy 8. Other policies are not optimal to be selected. The result also shows that when the rubber price is increased, policies 8 and 10 are even more attractive.

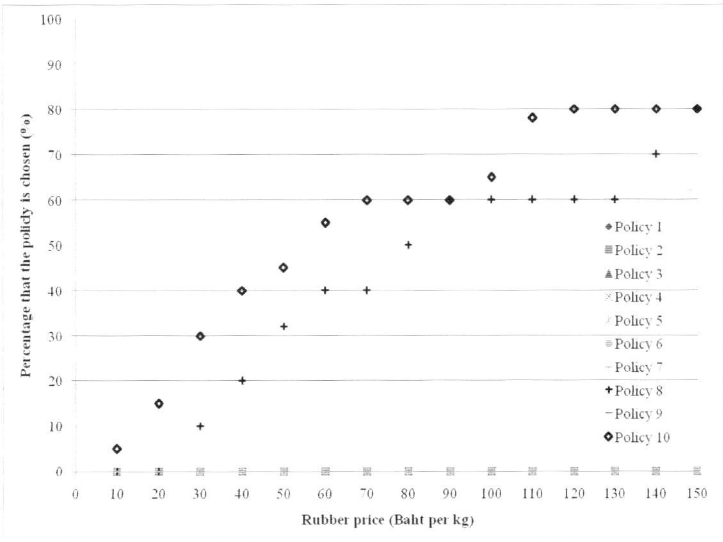

Figure 1: The percentage that the policies are chosen for each rubber price ranging from 10 to 150 Baht per kg

Another result also shows that when the rubber price is below 42 Baht per kg, policy 5 or 6 is also chosen to raise farmers' income so that their income passed the poverty line. Policies 1 and 2 are not optimal since these will only increase demand in a specific industry. If the benefit from policies 1 and 2 is increased (by increasing the demand significantly or by increasing the value of the final product), these two policies will then be attractive. Policy 3 is not optimal since the success probability to increase its demand is estimated to be between 10%-30%. If this success probability is increased to be more than 35%, then we can start to see the potential of this policy. Policy 9 is never selected since this does not affect the objective value of the policy maker (since the model is run for only one period).

4. Conclusion

The finding in this study provides a stepping stone for the Thai government in terms of the selection of rubber policies. Previously, the chosen policy has been mostly ad hoc based on the rubber price. Rubber farming policy selection should be more strategic in a long run. Strategic decision making should be a new trend of policy selection in Thailand, not only for rubber supply chain but also for any other agricultural product supply chain.

5. References

[1] Major J A 2002 Advanced techniques for modeling terrorism risk *J. of Risk Finance.* 4 1 15–24
[2] Zhuang J and Bier V M 2007 Balancing terrorism and natural disasters – defensive strategy with endogenous attacker effort *Operations Research.* 55 5 976–91
[3] Woo G 2002 Quantitative terrorism risk assessment *J. of Risk Finance.* 4 1 7–14
[4] Bier V M, Haphuriwat N, Menoyo J, Zimmerman R and Culpen A M 2008 Optimal resource allocation for defense of targets based on differing measures of attractiveness *Risk Analysis.* 28 3 763–70
[5] Haphuriwat N and Bier V M 2011 Trade-offs between target hardening and overarching protection *European Journal of Operational Research.* 213 1 320–28
[6] Azaiez M N and Bier V M 2006 Optimal resource allocation for security in reliability systems *European Journal of Operational Research.* 181 2 773–86
[7] Brown G, Carlyle M, Salmerón J and Wood K 2006 Defending critical infrastructure *Interfaces.* 36 6 530–44

IOP Conf. Series: Materials Science and Engineering **526** (2019) 012044 doi:10.1088/1757-899X/526/1/012044

Identification of bacteria caused bad odors in the rubber plants

K Choowongkomon[1], S Pumloifa[2], N Hansupalak[3], W Insuan[4]

[1]Department of Biochemistry, Faculty of Science, Kasetsart University, Bangkok 10900, Thailand 10900, E-mail: kiattawee.c@ku.th
[2]Interdesciplinary of Genetic Engineering and Bioinformatics, Graduate School, Kasetsart University, Bangkok, Thailand 10900.
[3]Department of Chemical Engineering, Faculty of Engineering, Kasetsart University, Bangkok 10900, Thailand 10900.
[4]Department of Veterinary Technology, Faculty of Veterinary Technology, Kasetsart University, Bangkok 10900, Thailand 10900.

Abstract. The problem of odor from the rubber latex industry is a major problem in the rubber industry. Due to complaints about odors from the people in the industry areas. One cause of odor may be associated with the biodegradation process by bacteria such as carbohydrates, proteins, etc., which are present in the serum. And gas generation. At present, industrial plants use water-absorbing odor to reduce the intensity of odor, but it is not very good. Therefore, if you can eliminate the microorganisms that cause odors. It may help reduce the intensity of the smell. This research was conducted to isolate microorganisms from rubber plantations and rubber plant for odor analyzed use in GC-MS technique. Comparing the gas composition with the gas data obtained from the actual air sampling. Odor composition analysis It found that the rubber and lump rubber composed with Acetic acid, propanoic acid, 2-methyl-propanoic acid, butanoic acid, 3-methyl-butanoic acid, pentanoic acid.

1. Introduction

The rubber industry is an important industry for the Thailand economy. In contrast, the amount of natural rubber production. Thailand is likely to decline. Due to weather problems in the South, rain has increased. And one of the major problems is. The bad smell of water serum affects people close to the rubber industry. For this reason, it is a major obstacle to the rubber industry in Thailand. The smell may be related to the process of biodegradation, such as carbohydrates, proteins, etc. contained in serum and gas generation. The microbes that grow in the production process are bacteria. There are three groups of bacteria is Aerobic Bacteria, Anaerobic Bacteria and Facultative Anaerobic Bacteria The smell of this acidic sour taste. According to Hoven and colleagues, volatile fatty acids is Acetic acid, Propanoic acid, Butanoic acid, 2-Methylpropanoic acid, Pentanoic acid, isovaleric acid and Hexanoic acid (Hoven *et al.*, 2003). At present, industrial plants use water-absorbing odor to reduce the intensity of odor, but it is not very good. The composition of natural latex consists of small particles suspended in serum. The rubber is a hydrocarbon, which is polyisoprene is a part. The average molecular weight is about 106 grams per mole. (Gazeley, 1998) Rubber particles in the latex are about 30-40 percent by weight. In addition to water and rubber particles, there are also organic and inorganic minerals included. (Blackley, 1997) The composition of latex consist of. 33 percent dry matter, 33 percent protein, 1 percent protein, 1-2.5 percent ash, 1 percent sugar, 1 percent sugar and 1 percent water (Blackley, 1997).

2. Methods

2.1 Sampling of bacteria and air samples

The researchers collected samples of microorganisms by swab and air samples from rubber plantations and rubber plantations. Sri Trang Agro-Industry Plc. Located in Ubon Ratchathani. Take the serial dilution sample to determine the appropriate concentration of each sample so that the number of colonies in the range of 30-300 cfu / ml. Dilution at 10^{-1}, 10^{-2},10^{-3} and 10^{-4}. Then cultured on a Nutrient Agar Plate of 1 ml culture overnight at 37 degrees Celsius, select different colonies in each sample, then bring the Streak on the Nutrient Agar Plate overnight at room temperature. 37 degrees Celsius. Choose 1 colonies to repeat the streak plate. (Tassadaq *et al.*, 2013) Finally a single colonized broth was fed to Nutrient Broth (NB) 3 ml for Gram straining.

2.2 PCR amplification

Pure bacteria are extracted from DNA for amplified 16s rRNA gene by Polymerase Chain Reaction. In this experiment the universal primer was used. Forward primer: 27f (5′AGAGTTTGATCMTGGCTCAG-3′)Reverse primer:1492r(5′TACGGYTACCTTGTTACGACTT-3′) Then analyzed the PCR product with an agarose gel electrophoresis the PCR product is purified to deliver DNA sequencing.

2.3 phylogenetic analysis

Bacterial nucleotide sequences were compared with online databases (GenBank) by alignment software (CLUSTAL W) and generated phylogenetic tree by MEGA V.7 program. The algorithm is defined as follows. Neighbour-Joining tree, Bootstrap Replications 1000, Model/Method Kimura 2-parameter model.

2.4 Preparation Selective media and Selection of Bacteria

Prepare the Selective Media, which consists of Rubber and Agar in rubber ratio: Agar is 1: 5. Then purified bacteria known to the Streak plate on Selective media incubate overnight. Transfer the colonies from each agar plate to a 1.5 ml centrifuge tube containing 1 ml of liquid containing 40% glycerol in Nutrient broth.

2.5 In vitro test generate odor from bacteria

Remove the rubber from the rubber plantation, cut into small pieces of 0.5 cm3. The rubber is put in a test tube of 3 g sterilized (sterilization is at 121 degrees Celsius 15 pounds per square inch for 15 min) Then add the above-described bacteria (500 µl volume at 8 Mcfaland) to a sterile tube, add sterile water to the test tube (control). Incubate at room temperature for 1 day, 3 days and 5 days and analysed by Sniffing.

2.6 Analysis of air samples by GC-MS technique.

Collection of air samples by comparing two methods of air sampling based on the National Institute for Occupational Safety and Health (NIOSH) and NIOSH Manual of Analytical Methods. Collect air samples in a 200 µl bag of gas and SOLID SORBENT TUBE (coconut shell charcoal, 100 mg/50 mg). That is, the adsorbent used for evaporating acid is acetic. Finally, analysed the composition of the gas by GC-MS technique.

3. Results and discussion

3.1 Identified Bacteria

The microbial isolation by doing serial dilution and then implied 16s rRNA gene fragment. Found the PCR product size analyzed by agarose gel electrophoresis approximately 1,465 base pair Figure 1.

Subsequent transmission of the nucleotide sequence by comparing the nucleotide sequence with the NCBI database. We found 39 species of bacteria.

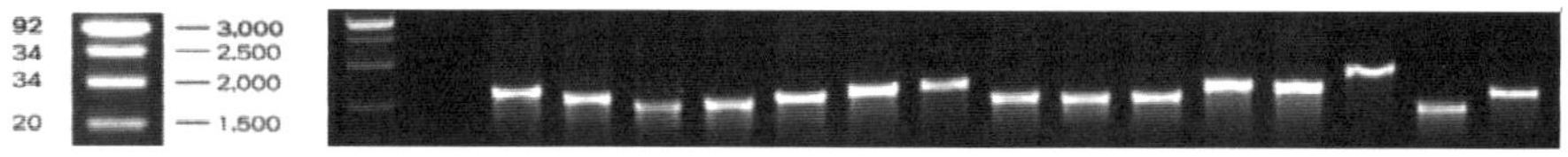

Figure 1. The PCR product size approximately 1,465 base pair

3.2 Selection of Bacteria from Selective media

Selective media is prepared from Rubber and Agar in rubber ratio: Agar is 1: 5. The results of all 39 species tested in previous studies were isolated for the growth of good bacteria using only the nutrients contained within the latex and the rubber. It found that the number of bacteria that can grow to only 18 species in Figure 2.

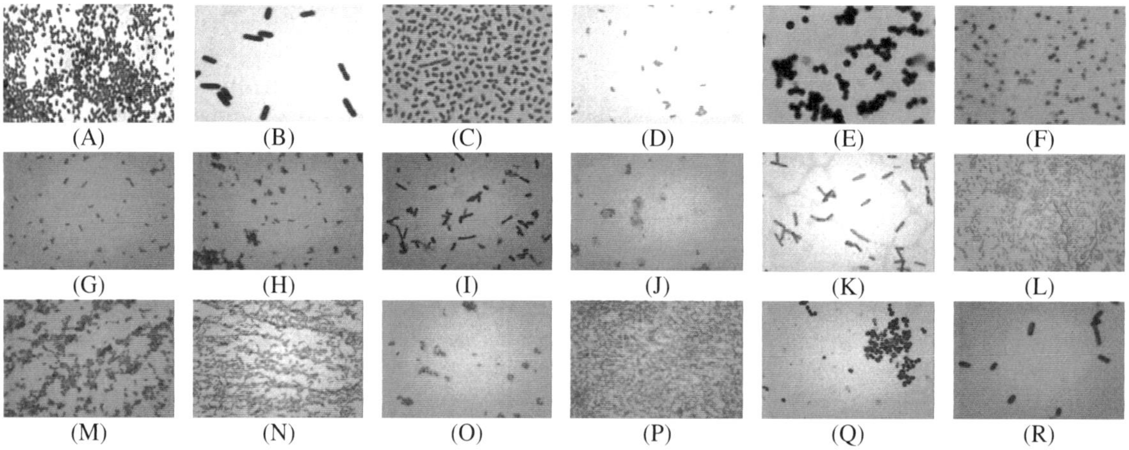

Figure 2. Bacteria gram stain; (A) Acinetobacter pittii, (B) Bacillus thuringiensis, (C) Cedecea davisa, (D) Citrobacter freundii, (E) Enterococcus faecalis, (F) Enterococcus feacium, (G) Enterobacter, (H) Klebsiella, (I) Kurthia gibsonii, (J) Lactococcus lactis, (K) Lysinibacillus xylanilyticus, (L) Myroides injenensis, (M) Pantoea wallisii, (N) Proteus mirabilis, (O) Proteus vulgaris, (P) Pseudomonas aeruginosa, (Q) Staphylococcus sciuri, and (R) Stenotrophomonas maltophilia.

3.3 In vitro test generate odor

This test uses a cup of rubber from the factory. Cut to 0.5 cm3, then sterilize at 121 degrees Celsius 15 pounds per square inch for 15 minutes when the sample is put into the liquid. *Bacteria n*o growth was found in *nutrient broth*. The fluid is still clear as before the infection. While the control set is Non-sterile rubber When put in *nutrient broth*, it is seen that the liquid is opaque. Bacteria growing on all 18 selective media were added to sterile rubber in *nutrient broth incubate* of 1, 3 and 5 days. The results of the smell. Can be classified into three groups according to the foul Table 1.

Table 1. The odor of bacteria

Very foul smell	Moderate foul smell	Little foul smell
Cedecea davisea	*Bacillus thuringiensis*	*Acinetobacter pittii*
Citrobacter freundii	*Enterococcus faecalis*	*Myroides injenensis*
Enterobacter cloacae	*Enterococcus feacium*	*Pseudomonas aeruginosa*
Klebsiella pneumonia	*Lactococcus lactis*	*Staphylococcus sciuri*
Kurthia gibsonii	*Pantoea wallisii*	
Lysinibacillus xylanilyticus		
Proteus mirabillis		

IOP Conf. Series: Materials Science and Engineering **526** (2019) 012044 doi:10.1088/1757-899X/526/1/012044

Very foul smell	Moderate foul smell	Little foul smell
Proteus vulgaris		
Stenotrophomonas maltophilia		

3.4 GC-MS

The cups of the cups were analyzed for their aroma composition. As shown in Figure 3 six volatile fatty acids were found. Analysis of coconut shell charcoal samples was carried out by GC-MS. We found that the sample area in the production process contains acetic acid.

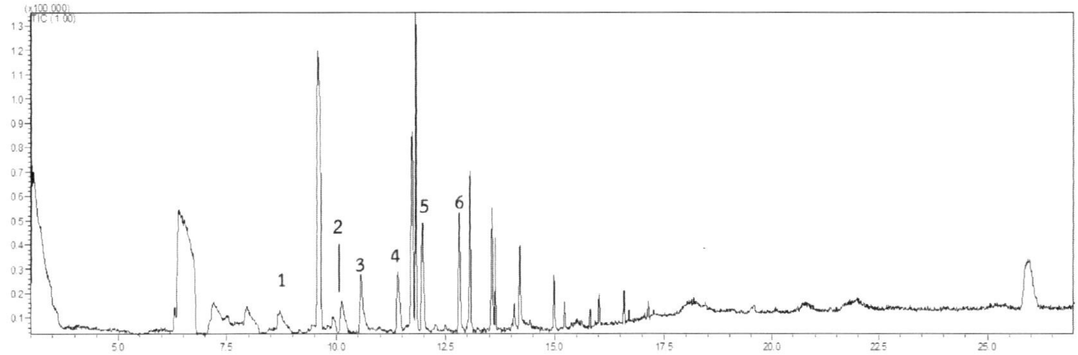

Figure 3. Show the chromatogram of the odor composition obtained from the analysis of rubber cups, *Number* 1 is acetic acid, number 2 is propanoic acid, number 3 is 2- methyl-propanoic acid, No. 4 is butanoic acid, No. 5 is 3-methyl-butanoic acid, No. 6 is pentanoic acid.

4. Conclusion

The samples were collected for selection of odor-causing bacteria in the rubber cup process. Bacteria were collected from rubber plantations and factories by selecting the bacteria from the actual sample by serial dilution and increasing the amount of genetic material to see the species. There are 39 different types of bacteria that can be isolated from selected rubber media at the ratio of latex: Agar (1: 5). Can grow in food prepared from Selective media. Inspect the odor content of the rubber cup by static headspace technique by slicing the rubber cup to a small size and filling it in a glass jar and heating it to make the vapors vapor. The air gap above the sample is injected into the analyzer by the technique. GC-MS found that VFAs were composed of acetic acid, propanoic acid, butanoic acid and pentanoic acid. According to Hoven and colleagues, volatile fatty acids, such as acetic acid, propanoic acid, butanoic acid, 2 -methylpropanoic acid, pentanoic acid, isovaleric acid.

5. References

[1] Blackley, D.C. (1997) Polymer Latices: Science and technology Volume 2: Types of lattices. 2ed. Vol. 2. Chapman & Hall, London.

[2] Gazeley, K.F., Gorton, A.D.T., Pendle, T.D. (1998) in Natural Rubber Science and Technology, A.D.Roberts (Ed.), Oxford University Press.

[3] Hoven, V.P., Rattanakarun, K., Tanaka, Y. (2003) Determination of chemical components that cause mal-odor from natural rubber, Rubber Chemistry And Technology. 76, 1128-44.

[4] Hoven, V.P., Rattanakarun, K., Tanaka, Y. (2004) Reduction of offensive odor from natural rubber by odor-reducing substances, Journal of Applied Polymer Science. 92, 2253–60.

[5] Tassadaq H., Aneela, R., Shehzad, M., Iftikhar, A., Jafar, K., Veronique, E.H., Kil, K.Y., Muhammad, A. (2013) Biochemical characterization and identification of bacterial strains isolated from drinking water sources of Kohat, Pakistan, African Journal of Microbiology Research. 16, 1579-90.

IOP Conf. Series: Materials Science and Engineering **526** (2019) 012045 doi:10.1088/1757-899X/526/1/012045

Surface treatment with microwave plasma improving PEDOT:PSS adhesion on natural rubber

K Meesawat[1], K Thana[2], R Jaisutti[2] and N Petchsang*[1,3]

[1] Department of Material Science, Faculty of Science, Kasetsart University, Bangkok 10900, Thailand.
[2] Department of Physics, Faculty of Science and Technology, Thammasat University, Pathumthani 12121, Thailand
[3] Specialized Center of Rubber and Polymer Materials for Agriculture and Industry (RPM), Faculty of Science, Kasetsart University, Bangkok 10900, Thailand

*Corresponding author: nattasamon.p@ku.ac.th

Abstract. Hydrophobic surface of pre-vulcanized natural rubber can be modified to a hydrophilic surface under plasma treatment. This is useful to make flexible and composite substrate from natural rubber in many applications. A low-cost microwave plasma cleaner was built using argon gas as a source for generating plasma. The plasma treatment conditions on rubber were investigated under two parameters: treatment time and treatment cycles. We found that the water contact angle was decreased when the pre-vulcanized natural rubber was treated more time or more cycles. To create the conductivity of natural rubber, poly(3,4-ethylenedioxythiophene) polystyrene sulfonate (PEDOT:PSS) in water was spin on the treated rubber surface. The electrical measurement shows conductivity on the rubber surface with a uniform current-voltage relationship. Without plasma treatment, PEDOT:PSS hardly attached on surface yielding no conductivity appeared on natural rubber.

1. Introduction

Natural rubber, now a day, is widely used in many applications [1]. It is excellent in high stretchable material which can reach high stress-strain compare to another polymer material [2]. In the form of pre-vulcanized latex, it is convenient to use because after heating the rubber show excellent stretching without destroying its structure [3]. Although natural rubber is a candidate for making flexible substrate in electronic devices, however, it will be more useful if it is used in the form of conductive rubber. To use rubber as a conductive flexible substrate, it needs another material to enhance its properties such as conductive polymer or carbon nanotube. Poly(3,4-ethylenedioxythiophene) polystyrene sulfonate (PEDOT:PSS) is a conductive polymer that is widely used in many application for electronics due to it has a good conductivity under stretchability [4]. To make conductive rubber, layer structure between rubber and PEDOT:PSS is convenient by using spin-coating method. However, the surface of rubber is hydrophobic while PEDOT:PSS is a solution in water. So after spin-coating, there is no PEDOT:PSS can be attached on rubber surface. This is due to high contact angle of rubber surface which prevent adhesion of water solution.

In this research, we would like to improve adhesion of PEDOT:PSS on rubber by reducing contact angle on the rubber surface. Plasma cleaning is appropriate and affordable method [5]. We construct a cost-effective plasma generator form a household microwave oven. After that, the contact angle of treated rubber was studied along with the treated cycles and treated interval time.

2. Methodology

2.1. Microwave Plasma system

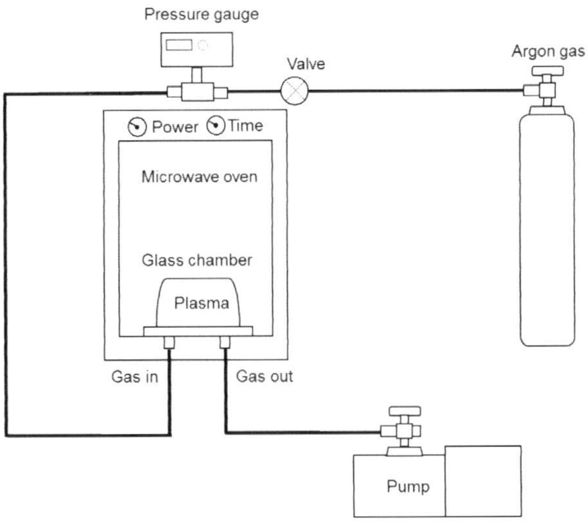

Figure 1. Schematic of microwave plasma system for cleaning rubber surface.

The plasma cleaner was built from a low-cost microwave oven as shown a schematic in Figure 1. A household microwave oven was modified and connected to the vacuum pump and argon gas tank. Inside the microwave chamber, a Pyrex bowl was inverted and placed on a rubber gasket. After putting the sample inside the chamber, the vacuum pump was run for removal of room air and humidity or water film on the sample surface. After that, the amount of argon gas was controlled by needle valves which were measured by a pressure gauge. The microwave oven was set to generate 800 watts of power continuously with adjusted time. The generated plasma bombarded onto the rubber surface. High heat will be generated during this operation. However, the rubber can be burnt if a longer time of treatment was applied to the sample.

2.2. Natural Rubber Cleaning and Coating Process

In this research, we used a pre-vulcanized rubber latex buying from a local company in Thailand. Its composition contains sulfur that is essential for making crosslink of rubber after baking. 15 mL of the latex was poured in a petri dish and baked at 70 °C for 3 h. After temperature goes down to room temperature, the rubber appeared light yellow color. After that, it was cut to 3x3 cm^2 and its surface was clean by distilling water followed by acetone in order to remove residues. Next, it was dry by using nitrogen gas from cylinder tank. The plasma system was then used to the deep clean rubber surface. Cycles of plasma were investigated by varying from 1 to 4 cycles. Treating time is also varied, these are 5s, 15s and 25s of each cycle. Then the contact angle of all samples was measured to evaluate the water drop angle on the treated surface. Finally, 400 μL of PEDOT:PSS with 1.1% concentration in H_2O was spin at the speed of 3000 rpm for all samples. The light blue color will appear after PEDOT:PSS attached on it.

3. Results and discussions

3.1. Water Contact Angle

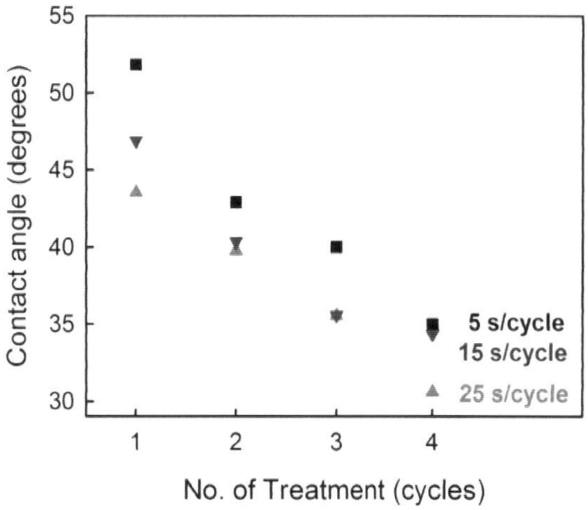

Figure 2. Relation of water contact angle of natural rubber and No. of plasma treatment of 5s/cycle,
15 s/cycle and 25 s/cycle.

The water contact angles of treated surface rubbers were measured by a digital camera for all samples. Before plasma treatment, the bare rubber showed average contact angle of 72.7°. With this high contact angle, after spin-coating, there is no PEDOT-PSS on the rubber surface. To reduce contact angle, plasma treatment was applied 1 to 4 cycles with 5s interval time. We repeated the same method with 15 s and 25s for comparison. Figure 2 showed the relation of contact angle and No. of treatment cycles for each interval times. On 5s interval, the contact angle appears 51.83°, 42.9°, 40.01° and 35.0° on 1, 2, 3 and 4 cycles, respectively. The result showed that when the rubber was treated with many cycles, the contact angle was reduced. This means that the plasma treatment can improve the contact angle of rubber surface. This is also similar when treating time interval change to 15s/cycle and 25s/cycle. On 15s interval, the contact angle showed 46.87°, 40.37°, 35.37° and 34.35° and for a 25s interval, it showed 43.55°, 39.77°, 35.55°, 30.61°. When the interval time of each cycle considered, we found that longer interval time yielding lower the contact angle.

Then all samples were coated with 1.1% PEDOT-PSS in water with spin-coating at 3000 rpm. The results showed that the light blue color appeared when contact angle lower than 36°. This occurred when the rubber was treated with 5s for 4 cycles, 15s for 3 cycles, 15s for 4 cycles, 25s for 3 cycles and 25s for 4 cycles. From these experiments, 25s for 4 cycles gave the lowest average contact angle. However, we selected the condition of treatment at 15s for 3 cycles since the adhesion of PEDOT-PSS appeared smoothly on rubber surface with shortest time and cycles.

3.2 Electrical Measurement

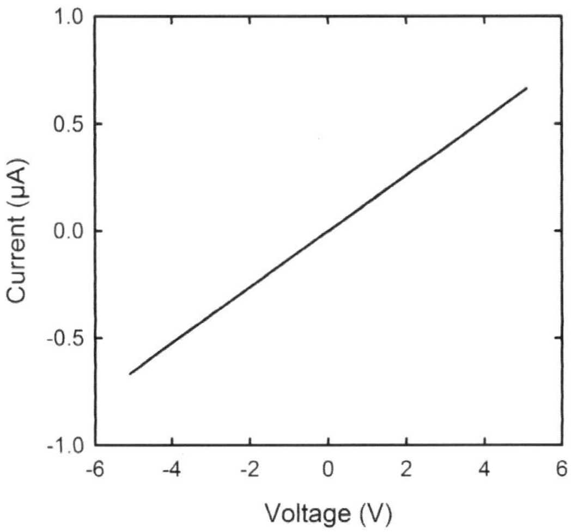

Figure 3. Electrical measurement of current-voltage of PEDOT:PSS on pre-vulcanized natural rubber.

To measure the electrical resistance of PEDOT:PSS on the rubber surface, the rubber that treated 3 cycles with 15s/cycle was then attached with a gold sheet which acts as an electrical contact for current-voltage measurement. The conductive rubber was then cut to 0.5 cm x 3 cm before voltage was applied on the cut rubber. The resistance of conductive rubber was extracted from the linear I-V curve which was ~2.7 MΩ/cm. The electrical measurement also showed a smooth coating of PEDOT:PSS on rubber surface as can be seen a slight distribution of electrical resistance.

4. Conclusions

A conductive natural rubber was fabricated from PEDOT:PSS coated on pre-vulcanized rubber. The contact angle of rubber was improved by treating with microwave plasma. The results showed that the contact angle was reduced after plasma treatment for 15s with 3 cycles. The contact angle showed that when it was lower than 36°, adhesion of PEDOT:PSS on rubber is higher. The electrical resistance on conductive rubber show ~2.7 MΩ/cm which is confirmed a good adhesive of PEDOT:PSS on rubber.

5. References

[1] Kaltseis R, Keplinger C, Koh S J A, Baumgartner R, Goh Y F, Ng W H, Alexander Kogler A, Tröls A, Choon Chiang Foo C C, Suo Z, Bauer S 2014 *RSC Adv.* **4** 27905-13.

[2] Chapman A V 2007 *24th International H.F. Mark-Symposium, 'Advances in the Field of Elastomers & Thermoplastic Elastomers'*

[3] Findik F, Yilmaz R, Köksal T 2004 *Materials & Design* 25 4 269-276.

[4] Hakimi M, Salehi A, Boroumand F A 2016 IEEE Sens. J. 16 6149–54.

[5] Bogaerts A, Neyts E, Gijbels R, Mullen J V D 2002 Spectrochimica Acta Part B 57 609–58.

Acknowledgements

This work was financially supported Thailand Research Fund (TRF) (Grant No. RDG60T0148).